Using the
Engineer's Lathe
in Clockmaking

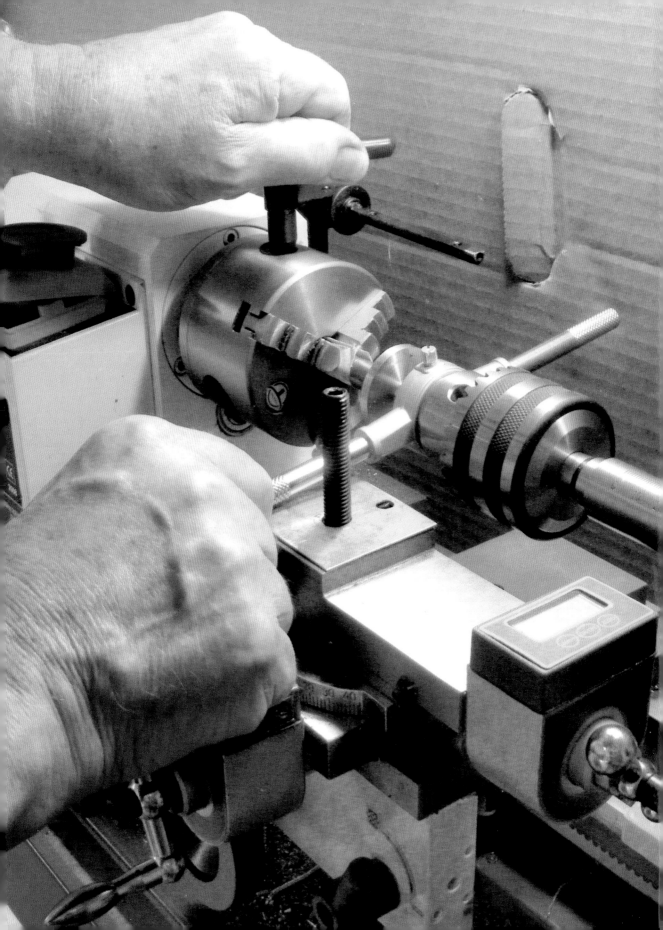

Using the Engineer's Lathe in Clockmaking

Laurie Penman

NAG Press

First published in 2022 by
NAG Press, an imprint of
The Crowood Press Ltd,
Ramsbury, Marlborough
Wiltshire SN8 2HR

enquiries@crowood.com

www.crowood.com

British Library Cataloguing-in-Publication Data
A catalogue record for this book is available from the British Library.

ISBN 978 0 7198 3151 5

Typeset by Jean Cussons Typesetting, Diss, Norfolk

Cover design by Maggie Mellett

Printed and bound in India by Replika Press Pvt Ltd

Contents

Acknowledgements

Fig. 1.01 by kind permission of Myford Ltd (www.myford.co.uk)

Figs 1.02 and 2.12 by kind permission of Chester Machine Tools (www.chesterhobbystore.com)

Fig. 1.03 by kind permission of Sherline Products (www.sherline.com)

Figs. 2.20; 3.12 and 4.04 by kind permission of Machine Mart Ltd (www.machinemart.co.uk)

Figs 4.12a, 4.12b, 5.06, 6.15 and 10.01 by kind permission of RDG Tools Ltd (www.machinemart.co.uk)

Chapter 1

Introduction

This book begins as an instructional manual, and consequently the first chapter assumes that the reader does not have a lathe at the moment and needs advice on choosing one. From there the chapters deal with preparing a cutting tool; the possible techniques; and several machining projects that are needed when repairing clock movements or that will prove useful clockmaking (or model-making) tools for your workshop. The intent is to provide a gentle learning curve for the practical use of the lathe.

CHOOSING A LATHE

Enjoy getting to know the engineer's lathe, as it is almost a universal tool. There are very few machining operations that it is not capable of – it is even possible, with a little cunning, to use it to make a larger lathe.

The small engineer's or centre lathes that are used for clockmaking and repair perform very differently from the much larger lathes used in general engineering, although their functions are much the same; however, general engineering makes far greater demands upon the machine.

Lathes such as the Myford (Fig. 1.01) have all the characteristics of a larger lathe: lateral adjustment on the tailstock, top slide, complete apron with traversing handle and screw-cutting controls and back gearing; while mini-lathes usually have a swivelling headstock for taper turning, traversing at the end of the bed, no

Fig. 1.01   A Myford lathe.

top slide and a fixed centre tailstock. Other differences result from the type and size of the materials that are machined.

Parts for clocks do not generally demand cutting tools with a wide variety of rakes and clearances for working within the limits of the tool

material. It is quite often the case that old-fashioned carbon steel tools are perfectly adequate for the job in hand. So long as the speed is kept down, a carbon steel tool is actually harder than most high-speed steels (HSS).

For convenience (and to avoid undue advertising), I will refer to 'small lathes' and 'mini-lathes', the first being simply a scaled-down engineering machine such as the Myford or the Chinese 7 × 12 (Fig. 1.02), and the second one with a swivelling headstock like the Sherline (Fig. 1.03).

So far as the basic clockmaker's or watchmaker's lathe is concerned, the only differences in operation fall under the heading of 'graving', that is, the use of a hand-held tool called a graver. Very often a top slide is added to the watchmaker's lathe and then it is

Fig. 1.02   *A Chinese 7 × 12 small lathe.*

Fig. 1.03   *A Sherline mini-lathe with a swivelling headstock.*

really a tiny engineer's lathe. Some reference will be made to this, but I cannot think of any operation that cannot be performed at least as well on the normal centre lathe (which is the general term for the type of lathe discussed in this book). Precision turning can be carried out with a graver, but it has limited applica-

tion to modern clockmaking. It must be almost impossible to use a graver to remove an amount smaller than 0.025mm from a turned diameter and leave a high finish, or to modify a taper by a similar amount. Clockmaking uses files and burnishers for these tasks and they can be used just as easily on the centre lathe as on the tradi-

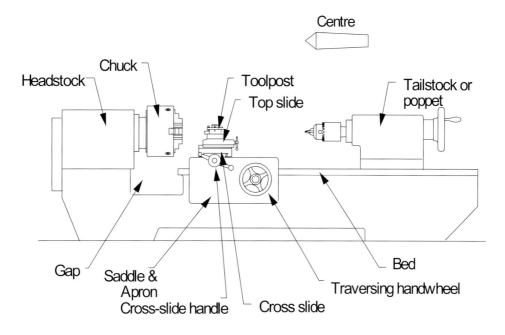

Fig. 1.04   The parts of a lathe.

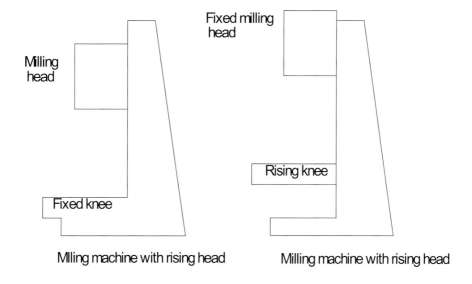

Fig. 1.05   The two types of milling machine: rising head, fixed knee (left) and fixed head, rising knee (right).

tional clockmakers' machine. The parts of the lathe are shown in Fig. 1.04.

Larger work, such as turning a bar larger than 25mm (1in) and longer than 300mm (12in); using a collet on a rod of 12mm (0.5in) diameter that passes right through the collet; the making of clock gears and tools or carrying out shaping and milling operations will all call for the larger machines. Gear-cutting requirements can be met on the Myford with attachments (either purchased, or made on the lathe), for gears up to about 200mm (8in) diameter. Many mini-lathes are sold with the choice of milling attachments and these will enable the cutting of gears up to about 75mm (3in) diameter, but the attachments are not as rigid as a purpose-made milling machine or wheel-cutting engine and will usually only have a small range of tooth counts available for the dividing device.

Vertical milling machines of solid construction, with a rising head rather than a rising 'knee' are available and not much more expensive than the mini-lathes. Wheel-cutting engines are even more useful for gear cutting and range in price from less than a vertical miller to just over. However, they will not carry out other milling operations. Milling machines with rising knees lift the work up to a rotating cutter; the rising head version has a static support for the work (with two slides) and the cutter is brought down to it. The support in small machines is cast as one with the frame of the machine and is much sturdier than the rising knee milling machine.

A professional clockmaking shop really needs all three of the machines mentioned (small lathe, mini-lathe, vertical miller or wheel-cutting engine) but the beginner will find that the mini is quite sufficient to begin with, progressing to larger machines if and when the work demands it.

Lathe Qualities

The major requirements of a good lathe are listed below:

Sufficient power   A useful rule of thumb for the light lathes used by clockmakers is 50 watts per inch of 'swing' (ignore any gap), for

speeds up to 2,000rpm. 'Swing' refers to the maximum diameter that may be accepted over the lathe bed. An 18cm (7in) lathe should have at least 350 watts (just under 0.5hp) and a 7.5cm (3in) lathe 100 watts. There are two electric motor 'ratings' – intermittent and continuous. Intermittent rating will require the motor to be switched off at frequent intervals so that it can cool off; continuous rating is self-explanatory: there is no need to switch off the motor.

A stiff bed   Every part of a lathe is related to the bed, and if it shows any tendency to distort under working conditions, the relative positions of headstock, tailstock and tool will change. This can result in 'chatter' or dimensional and geometric inaccuracy in the work piece. There is no way that you can test this but, as a rough rule, the width of the bed (the sliding surfaces) should be much the same as the height of the spindle over the bed. The higher the spindle (lathe arbor) in relation to the bed, the less stiff the machine becomes.

Accuracy   The importance of this very much depends on the type of work that the machine is required to do. If the lathe is to be used for boring out hydraulic cylinders, it would be expected to maintain a parallel cut over a longer distance than what is needed for clockmaking. In standard machine shop practice, the taper allowance on a 300mm-long turned bar is 0.025mm on the diameter and a concavity of 0.025mm on a 300mm diameter face. Practical requirements for clockmaking would be about 0.001in per inch for turning and 0.001in concavity on a 1in-diameter face (0.025mm per 2.5cm). This is from a new lathe, but experience will enable a turner to produce accurate work on a robust machine with much looser tolerances than this. Accuracy is very much a matter of how the work is tackled, but an accurate lathe (in the terms set out above for clock repairing) is easier to 'set up' than an inaccurate one.

Range of speeds   Speed changing can be managed either by means of pulleys, or gears from a constant-speed motor, or by using a variable-speed motor. A geared, or pulley-driven headstock, has the advantage of maintaining its speed well regardless of the size of cut (as long as the motor is powerful enough), but it has to work to a limited range of speeds. A variable-speed motor has an infinitely adjustable speed between a stated upper and lower limit, but the speed can alter as the amount of cut varies, or the state of the material being worked alters. An additional advantage not usually stated is that variable-speed motors can often be made to rotate at very low speeds indeed when there is no load on them. The ability to turn the chuck at 5 or 10rpm when setting work up is very convenient.

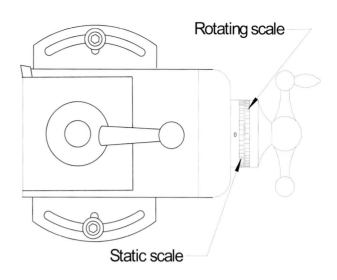

Fig. 1.06   *The scales are engraved with divisions of 0.025mm or 0.001in. The rotating scale is attached to the traversing screw and shows how far the tool has been moved.*

Convenient hand-wheels   The hand-wheels for moving the slides should be large enough to get one's fingers on and operate smoothly. An otherwise good lathe can be spoilt by this lack, and it would be a very good idea to increase the size of hand-wheel by making another that attaches to it or replaces it. Much the same applies to the tailstock handle. This is only an issue with mini-lathes in the main. Scaled dials should always be a part of the handles, marked clearly in either 0.025mm or 0.001in divisions. Some machines provide an adjustable scale ring that can be moved to zero and then locked in position, but on a small or mini-lathe the attachment tends to be a little unreliable.

Headstock rigidity It should be possible to turn a steel rod that protrudes from the chuck four times as long as the diameter of the piece without the work riding up and over the tool, or chattering. (The work piece must be stout enough not to bend and chatter on its own account, which is why its length is quoted its length in terms of its diameter.) This is quite a normal requirement in model- and clockmaking.

Sturdy head bearings   The arbor may be supported by a variety of bearings – ball, roller and solid. I prefer the last two because they tend to last longer in service. For non-industrial turning (clock repairing, for instance) solid

bearings have the great advantage that when, after a decade or two, they are too worn for good work they can be faked up for one last job – turning a new set of head bearings for the same lathe. The sturdiness of the headstock may be tested by mounting the largest diameter of steel or brass bar that will pass through the mandrel (spindle) and gripping it with the chuck before straining it manually up and down – in other words, trying to wiggle it.

Economic Considerations

The choice between a small lathe and a mini will depend on the cash available and the type of work that it is proposed to tackle. Myford lathes are no longer being made but a refurbished and guaranteed model with motor, three-jaw chuck, tailstock chuck, hard centre and a couple of high-speed tool pieces (the minimum 'ready to use' specification), would be around £1,500. A mini-lathe with the same accessories costs between £300 and £450 (at 2018 prices) but is much more limited in what it can do. Second-hand small lathes are more likely to be in good condition than second-hand minis, because they are much more robust machines.

If the work is to be limited to repair techniques – polishing, pivoting, bushing and so on – then a good mini-lathe will be quite sufficient.

Pros and Cons of Typical Lathes

Sherline and Similar 90mm (3.5in) Swing Lathes

Pros: High speed; sturdy; suitable for machining bar stock up to about 12mm (0.5in) diameter or short pieces and light turning of possibly 25mm (1in) and a disk held on a mandrel up to about 65mm (2.5in) safely. In particular, the Sherline has a very large range of add-ons, such as dividing head, quick-change tool posts, CNC and so on, making it very suitable for making and repairing many domestic clock parts. It's also inexpensive.

Cons: This is a small machine and cannot be used to repair spring barrels, for instance.

Myford

Pros: A big machine and capable of producing any part of a domestic clock, even gear wheels for tower clocks. There are very many extras, such as vertical slides, milling attachments, dividing heads, quick-change tool posts and CNC.

Cons: An excellent machine but over the top for simple clock repair tasks; it is more useful for clockmaking but expensive.

7 × 12 Chinese Lathes

Pros: These are inexpensive, simple, basic machines with a good range of speeds (infinitely variable) and sturdy; suitable for clock repair and making. Quick-change tool posts are available.

Cons: Not easily adapted for gear cutting.

LATHE OPERATIONS

The main operations will fall under the headings turning, boring, facing and milling (screw cutting and honing will be dealt with separately).

Turning   The action of producing a cylinder by rotating the work and holding a single pointed tool against the outside. Without the use of specialist devices, the effect is always a truly circular cross-section – unless there is a fault in the machining method. Sometimes it is convenient to use a flat file instead of a turning tool; early clockmakers frequently used a file for arbors, posts and pins.

Facing   Machining surfaces that are at right angles to the axis of the lathe.

Boring   The production of circular holes by rotating the work piece and holding a single pointed tool against the inside circumference.

Drilling   Boring cylindrical holes by holding a twist drill (or any form of drill) in either the tailstock chuck or the headstock chuck.

Milling   The production of machined surfaces using a rotating tool with one or more cutting faces. The work is usually not rotated and is either advanced in one dimension against the cutting action of the tool, or the tool itself is advanced in one plane to produce the same effect. The lathe may be used for milling operations without modifying it by holding work on the cross slide and mounting a milling cutter (end mill or slot drill) in the chuck. Movement of the cutting tool in more than two planes is unusual in clockmaking or repairing operations.

Chapter 2

Turning

Turning for most clockmaking tasks does not call for the same variety of tool forms that are needed in industry; clock repair machining (where production speed is of less importance than quality and cost of tools) rarely requires any more exotic material for the cutting tool than high-speed steel (HSS). I shall, however, briefly touch on other cutting materials because there are occasions when only a special tool will solve a machining problem.

A tool for plain, or cylindrical, turning has:

- A side rake (or top rake) to clear the metal being removed and provide one plane of a cutting edge
- Side relief to prevent the tool from rubbing on the work and to provide the other plane to the cutting edge
- Front relief, to avoid rubbing the newly exposed surface of the work.

Table 1 shows typical angles for these three faces when turning free-cutting (machining) brass, mild steel and high-carbon steel (silver steel or drill rod). Note that these angles are for work that is demanding; however, most small work (such as clock parts) can make use of the same angles, simply modifying the turning speed to prevent the tool from overheating. This makes for a great deal of economy in tool usage. The angle is about 10 degrees for the top clearance and 5–10 degrees for the clearance of front and side. There is rarely any need for rake.

These are guidelines only, as the angles for front and side clearances affect the support to the cutting edge and point. The tougher the work, the smaller this angle should be. The top rake affects the sharpness of the tool, the speed at which waste metal (swarf) is cleared away from the work piece, the mass of metal behind the cutting edge and consequently the rate at which heat is drawn away from it.

Table 1   Typical rake/clearance Angles for turning brass, mild steel and high-carbon steel

Tool	Cast iron	70/30 brass	Machining brass	Mild steel	Silver steel
Carbon steel top	10/15	10/15	5/10	10/15	5/10
Carbon steel side	10/15	10/15	5/10	10/15	10/15
Carbon steel front	10/15	10/15	5/10	5/10	5/10
HSS top	10/15	10/15	5/10	10/15	5/10
HSS side	10/15	10/15	5/10	5/10	5/10
HSS front	10/15	10/15	5/10	5/10	5/10
Tungsten carbide top	5	5	5	5	5
Tungsten carbide side	5	5	5	5	5
Tungsten carbide front	5	5	5	5	5

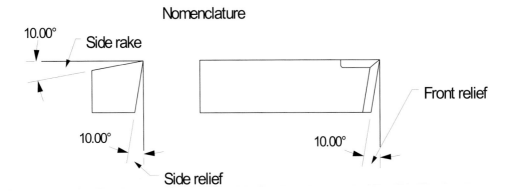

For light turning and facing the clearances may be the same for brass and steel

Fig. 2.01   *The terms used to describe the features of a tool for plain or cylindrical, turning.*

METALS

70/30 brass   The numbers indicate a metal composed of approximately 70 per cent copper and 30 per cent zinc. This type of brass is used for hammering or 'drawing' on a press; it is not good for machining. When turned, the metal produces long, curling swarf (the metal equivalent of wood shavings). In clockmaking it is mainly used sheet as sheet and employed for parts like lifting pieces, rack hooks and so on. It becomes springy if hammered and it is brittle when red hot (hot short). Cast parts are often made from this alloy, and before the mid-eighteenth century all brass parts were 70/30 with a small amount of lead.

Free-cutting brass (machining brass) The proportion of copper to zinc in this alloy is approximately 65:35. It is a brass that machines easily and is used for pipe fittings, clock wheels, collars and mountings. The swarf sprays off the tool in short chippings. This alloy tends to crack when bent cold but is readily deformed when red hot (cold short). It is also used for forgings.

Mild steel   This low-carbon steel that does not harden appreciably with heat treatment. It can however be case-hardened where a thin stratum of carbon-rich steel is produced by a heat treatment. The stratum is really thin and,

when hard surfaces are required, high-carbon steel is more reliable. Mild steel is used for arbors and semi-hard parts. There is a certain amount of hardening during 'working' (work hardening) and burnishing is reputed to provide a thin, hard surface to pivots.

Silver steel or drill rod (high-carbon steel)   This is used for arbors that have hard parts (pivots), or parts that are subject to wear such as pinions and escapement pallets. Tool steel has a higher carbon content (with other elements) and can be used for making tools that cut or punch. As its name implies, it is used more for making tools than clock parts.

HSS   This steel retains its cutting edge at higher temperatures than silver steel or drill rod and is commonly used for light turning jobs – the main use in clockmaking.

Tungsten carbide   This very hard material requires a green-grade grinding wheel or diamond to sharpen it. It is rarely needed for clockmaking tasks.

BASIC TURNING TOOLS

Fig. 2.02 shows a tool for turning the pivots, arbors and spindles up to about 12.5mm in

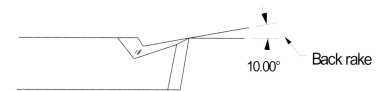

For heavy turning a top relief is needed and that and clearance will vary for the material being machined and the duty required. Only applies to 7 inch lathes and larger.

10.00°   Back rake

Fig. 2.02   This is a tool for turning the diameter of pivots, arbors and spindles up to about 12.5mm (0.5in) in diameter.

diameter. This is not the upper limit of the metal that can be machined, but above this diameter it will be found that the speed of the metal over the tool will require light cuts in comparison to the diameter. A cut of 0.25mm on a steel bar of 3mm diameter (a pivot, for example) is a moderately heavy cut and there will be a tendency for the bar to bounce unless it is short in length

(less than four times the finished diameter) but the same amount removed from a 19mm-diameter bar is a very light cut. As far as the tool is concerned, however, the amount of work being done is the same.

This tool can be used for roughing out small work and then, after touching up the cutting edge (if necessary) with a carborundum stone

Fig. 2.03 A light cut taken along a machined bar indicates whether it is running 'true' or not.

and by swivelling it to the position shown in Fig. 2.04, it may be used for finishing the diameter and the shoulder. It will also provide a slight undercut at the angle of the shoulder. I prefer the tool to be as keen as a graver when finishing diameters like those of a pivot. This will make the use of a fine Arkansas stone to fin-

ish the faces imperative. Test the keenness of the point by holding it loosely between thumb and forefinger and dragging it across your other thumbnail. If it slides it is not sharp enough; if it catches then it is just right. Finally, use the Arkansas stone to slightly dull the vertical edge by no more than a hair's breadth.

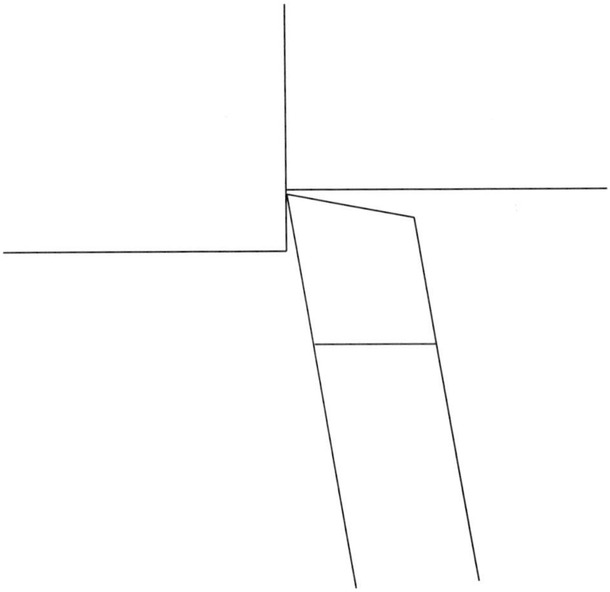

A glance at any of the books on commercial turning will show that the forms of turning tools are nearly all as shown in Fig. 2.06: the cutting edge slopes to the right and there is an appreciable chamfer or radius at the tip – this is to prevent the tip from being broken or worn easily. By comparison, the tools that I use mostly have very little work to do and I am more intent upon producing faces and shoulders that make a right angle to the long axis of the work.

When grinding a stick of HSS to make a new turning tool, the top rake is ground without touching the side surface and a 'witness' of 0.1mm is left along the left side of this top face. When the side relief is ground, a similar witness is left on the side relief or upright face. The tool is finished with a carborundum or emery stone to remove the witness and produce a sharp edge that as near as possible is the original edge of the HSS stick. When the tool needs sharpening, only the front needs to be ground away to expose fresh cutting edges. I used to have tools in my box that had been used for more than ten years without either of the relief surfaces needing to be refreshed.

Fig. 2.04   Diagram showing the ideal position of the tool against a carborundum stone to finish the diameter and the shoulder.

Fig. 2.05   Fine Arkansas stone is used to slightly dull the vertical edge.

Fig. 2.06   Standard turning tools: the cutting edge slopes to the right and there is a chamfer or radius at the tip.

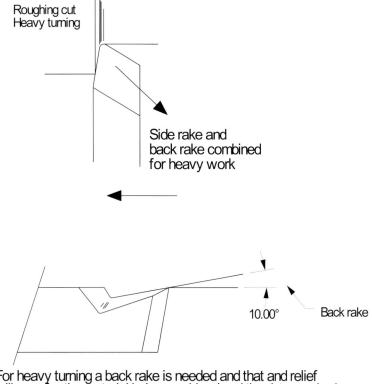

Roughing cut
Heavy turning

Side rake and back rake combined for heavy work

10.00°   Back rake

For heavy turning a back rake is needed and that and relief will vary for the material being machined and the duty required. Only applies to 7 inch lathes and larger.

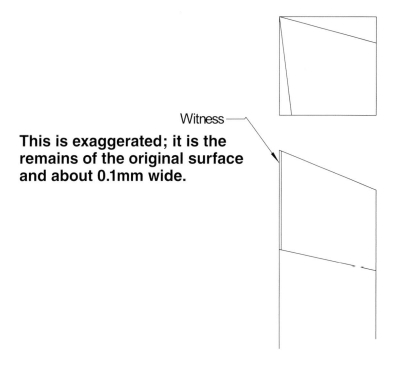

Witness

This is exaggerated; it is the remains of the original surface and about 0.1mm wide.

Fig. 2.07   The top rake is ground without touching the side surface, leaving a 'witness' of 0.1mm along the left side of the top face.

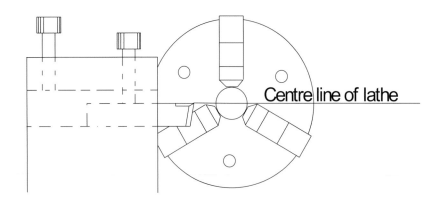

Fig. 2.08   Ensure that the top surface of the stick of HSS lies on the centre line of the lathe.

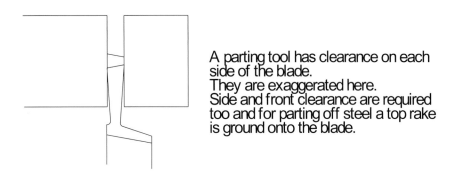

A parting tool has clearance on each side of the blade.
They are exaggerated here.
Side and front clearance are required too and for parting off steel a top rake is ground onto the blade.

Fig. 2.09   A parting tool is used to either cut the work from the metal bar or to make a channel.

The dimensions of the stick of HSS should be chosen so that when held in a square tool post, the top surface lies on the centre line of the lathe. Most lathes will be produced with a particular size of stick in mind. If this is not the case for your lathe, make a packing piece for the size of HSS that you intend to use. This is simply a piece of metal that is machined or filed to a thickness that keeps the top edge of the HSS level with the lathe's centre line.

Fig. 2.09 shows a parting tool, which is used either to cut the work from the metal bar or make a channel. The front of this is angled so that when cutting off the finished work, there is a very small 'pip' left at the break-off. If the tool is to be used for cutting grooves for circlips, keyhole washers and the like (or even as a help in left-hand turning), the front edge is best left square.

Turning the right-hand side of a bar, with the tool moving from right to left, is termed right-hand turning, and all tools used on this surface are also 'right-hand'. Turning in the opposite direction, with a larger diameter on the right, is left-hand turning. This all seems fairly obvious, but it is not an international understanding; for example, pivot files made in Switzerland for the right hand are for some reason called left-hand. If a tool is to be purchased in the ground condition, play safe and enclose a sketch of the form expected.

The size of the metal used for tools is not

*Fig. 2.10   Left-hand
turning, with the tool moving
from left to right.*

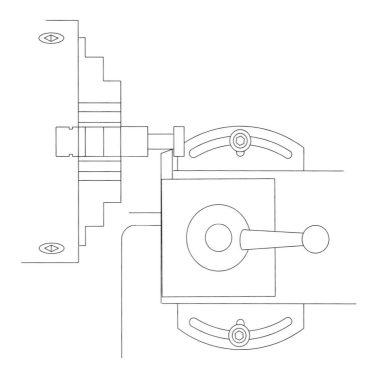

solely a matter of the strength required. Since
successful cutting requires the tool to be set
to the centre line of the lathe, it is very useful
to ensure that the tool remains on this centre
regardless of the need to resharpen – and here
again advantage can be taken of the smallness of
the work to veer away from standard engineering practice.

The normal side rake on a general engineering tool is as was shown in Fig. 2.06. It comes
back at an angle to the diameter of the turned
bar and helps to clear the swarf away from the
freshly turned surface. This is termed
side rake, as illustrated at the start of this
chapter in Fig. 2.01. The side elevation
shows the result of this. In small turning,
the slope can be kept normal to the side
of the tool, making it side rake only, and
then the tool may be sharpened by grinding the front face alone. In this way the
height of the cutting surface will remain
constant throughout the life of the cutting

edge. Light work of this type rarely pits the top
surface of the tool and an occasional rub with a
carborundum stone should keep it smooth and
remove any metal that adheres to the top.

Filing

A file can be very useful for certain tasks, such
as producing long, thin spindles like the arbor
extension for the minute hand. Filing on the
lathe is often scorned, but many of the antique
clocks have arbors and other parts that were

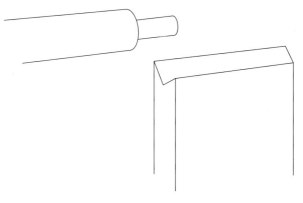

*Fig. 2.11   A simple method of filing using a
V-shaped holder in a vice and a file.*

filed originally – often not even in a lathe, but simply in a vice using a piece of scrap brass with a groove made in the end to support the steel while it was twiddled between finger and thumb as a file was stroked across it with the other hand.

In order to produce a reasonable cylinder, you will need one medium-cut flat file with smooth sides and another with a smooth side (teeth ground away) with good cutting edges to file the shoulders of the pivot. After producing a seating for the pinion head, hold the outside diameter of the pinion in the chuck and turn and polish a new pivot (*see* Chapter 8).

When using a file, always remember these safety rules:

- Hold the handle of the file in the left hand and do not reach over the revolving chuck.
- Roll up your sleeves or make sure that the cuffs fit the wrist closely.
- Make sure that the handle of the file is secure and not likely to expose the pointed tang of the file.

TOOLS FOR DIFFERENT MATERIALS

The intention behind the design of any cutting tool is to provide a sharp cutting edge supported by as small a mass of metal as can be managed, because the bulk of supporting material interferes with efficient cutting. A knife for cutting soft tissue will probably have an included angle of 20 degrees and it does not develop heat at the edge. It therefore needs very little mechanical support or conductive mass to remove heat.

Chisels for wood turning require more support to the edge and also need a mass of metal behind the edge to take heat away rapidly and prevent the tool from losing its hardness. However, too much mass will make the cutting edge less effective, and so wood chisels and plane blades typically have a cutting edge with an included angle of 30 degrees.

Metal turning puts a greater load on the edge and generates more heat. The harder the material to be turned, the more support must be given to the working edge and the greater the amount of heat that must be taken away by the tool or a coolant. Decreasing the angles forming the cutting edge increases the mass behind it.

Carbon steel rapidly softens at temperatures of 250°C and more. High-speed steels (HSS) will hold their hardness up to about 330°C. Tungsten carbide retains its hardness at very elevated temperatures indeed, but its effectiveness is limited by the material that holds the particles of carbide together. These can be designed to withstand red heat. Correctly designed tungsten carbide tools can remove metal that is glowing brilliantly and yet maintain their cutting edge. It is a little superfluous to point out that this is not necessary for the machining involved in clock repairing.

The first two metals are true alloys, the third consists of particles of the hard carbide bound into a softer matrix (often mostly nickel) and it demands a different treatment of its cutting planes. High-speed steels and carbon steels (silver steel) can be made to the same tool form as each other but will be used at different speeds. Tungsten carbide tools have to be formed to allow for the fact that they fail as a result of pitting behind the cutting edge, which weakens

Table 2   Cutting speeds in metres/minute (feet/minute)

Tool material	Cast iron	70/30 brass	Free-cutting brass	Mild steel	Silver steel
Carbon steel	15 (50)	30 (100)	33 (120)	18 (60)	15 (50)
HSS	25 (80)	60 (200)	76 (250)	37 (120	25 (80)
TC	76 (250)	170 (550)	275 (900)	91 (300)	76 (250)

the tool and allows the still serviceable particles of carbide to be pulled loose by the swarf. Tungsten carbide tools generally have rakes and clearances about two-thirds of those for HSS, or even less. (Negative rake is quite usual in carbide tools when the machine is robust enough to take advantage of the method.)

Table 1 showed angles common for the three cutting materials. Table 2 shows cutting speeds in feet per minute and metres per minute; the machinist has to convert this to revolutions per minute for the diameter that they are machining.

Feed

The depth of cut and the amount of 'feed' (the speed at which the tool progresses along the work piece) affect the speed of cutting; the values in Table 2 are for the typically light cuts associated with small turning. Most lathes that are fitted with an infinitely variable motor do not carry any means of measuring the speed. Only experience with your own machine will enable you to machine at the optimum speed each time, but if you start at a speed that is clearly lower than that needed and then speed up until the tool cuts smoothly, without 'chatter' and without overheating (the tool will discolour and so will the metal removed) at the depth of cut being taken, then the speed will be correct.

This may sound rather like being advised to get out of the car just before the accident, but it is possible to judge whether the tool is getting close to its limiting temperature. A carbon-steel tool should not heat to the point where its colour passes through light amber towards dark amber. High-speed steel can be taken to dark amber but not blue. It is always better to avoid colouring the metal at all.

With known pulley or gear ratios in the headstock and the stated speed of the motor (see the label that should be pinned to the motor frame giving voltage and other values), it is quite easy to calculate the speed of revolution of the work piece. The required revolutions will be the recommended speed for the material, divided by the circumference of the turned section. I have used feet and inches for this example. The cutting speed for mild steel, using a high-speed

steel tool is 120ft/min (37m/min). If the diameter turned is 1.25in (3.18mm), the circumference will be 3.927in (99.7mm).

$$\text{Revs} = \frac{120 \times 12}{3.927} = 367\text{rpm (revolutions per minute)}$$

The nearest available headstock speed is chosen and may well be 400rpm. Unless a very strictly governed workshop is involved, with close control of the material, state of tool and machine, this is as close as one needs to calculate. Any modification that is needed will be made apparent by the tool or the swarf discolouring, or the work piece chattering or otherwise objecting to the demands being made of it.

In practice, a clockmaker will most frequently be dealing with steel that is less than 0.25in (6mm) in diameter and 0.785in (19.9mm) or less in circumference. Then the calculation would be:

$$\text{Revs} = \frac{120 \times 12}{0.785} = 1,834\text{rpm}$$

A lathe that can rotate at 2,000rpm is indicated.

Making or restoring pivots that are no larger than 0.062in (1.6mm) will be a frequent task – at a recommended speed of 7,500rpm. It is clear that when all other requirements have been met, the final arbiter when you are choosing a lathe to buy is the maximum speed available. What is the top speed? In all probability it will be about 2,500rpm!

The Sherline (Fig. 1.03) has an infinitely variable speed range of 70–2,800 rpm; the Myford ML7 (Fig. 1.01) has a range of 35–640rpm, and the much more expensive Super 7 has an upper limit of 1,200rpm. The 7 × 12 Chinese lathe (Fig. 1.02) has speeds from 50 to 2,500 and is infinitely variable.

Judging Speeds and Depth of Cut

Calculations of feeds and speeds are very much a matter of providing guidelines only for the amateur or occasional machinist. (They become important when the judgement of the operator is removed, as in automatic machining or computer aided machining.) Since most turning in clockmaking is done over short distances, traversing automatically by powering the lead screws for the saddle or cross slides is risky and unnecessary. Manually advancing the tool is safer and your eye – and the machine – will tell you if the tool is being advanced too quickly. If the colour of the swarf is purple (brass), blue (steel) or the tip of the tool is coloured then the cutting edge is being overloaded or it is traversing too quickly.

The depth of cut can be judged by the tool chattering or the machine slowing down and 'protesting'.

SHARPENING TOOLS

The beauty of a high-carbon steel turning tool is that it can be filed to shape in the soft condition and then hardened and stoned to a fine finish. This is a very rapid way of providing a special tool. It is unnecessary to use a bench grinder; simply finishing the surfaces after hardening and tempering with a fine emery stone and an Arkansas stone is sufficient.

High-speed steel is much more difficult to

Fig. 2.13   Goggles or a full face mask should always be worn.

soften or harden without the use of an accurate means of measuring temperature and a precise knowledge of its constituents. Unless you have the facilities of a metallurgical laboratory, do not attempt it.

It is almost impossible to manage without a grinder of some description; stoning (using a small carborundum or other abrasive by hand) is very slow if there is any amount of metal to remove. A good 100 or 150mm-diameter (4in or 6in) grinder with the motor mounted directly on the shaft is a compact and quiet way of sharpening your tools. Fig. 2.12 shows a typical example.

When buying a grinding wheel, always state the metal that is to be ground and the speed of the grinding machine. Abrasives vary in grit size and strength; size of voids between the grits; and toughness of the bonding material. The retailer

Fig. 2.12   A 150mm (6in) bench grinder.

should have a comprehensive chart of characteristics provided by the manufacturers and is able to supply the correct wheel for your needs. Grinding a soft metal such as mild steel or brass on a wheel intended to grind HSS will clog the voids between particles, damage the wheel and produce a lot of heat.

Whatever means of grinding are available, follow these rules:

- Ensure that the grinding machine has a firm support for the tool to rest on and which will present the metal at the required angle; it must not slip while in use.
- At least three-quarters of the wheel circumference should be guarded with a solid metal case.
- There must be a transparent shield over the rest of the wheel so that the operator is protected from pieces of metal flung off the stone.
- Wear goggles or a full face mask to protect the eyes from particles that bounce from the work or any part of the machine. A full face mask (Fig. 2.13) will protect the skin of your face as well – it is quite possible for particles to be travelling so fast that they embed themselves in skin.

That is a long catalogue of requirements, but grinding is dangerous without all the safety pre-cautions. A piece of grit torn from the wheel can enter the eye at a speed of approximately 80km/h (50mph).

The actual grinding operation develops heat in the tool; if this is allowed to raise the temperature to dark amber or blue, the edge of a high-speed steel tool will be ruined, and the metal that is just behind the edge also. Always grind as coolly as possible. If a simple bench grinder is used, the tool must be removed from the wheel as soon as it becomes uncomfortably hot to touch at the ground end and dipped into a container of water to cool. More expensive machines are equipped with a coolant (usually clear water) that runs over the wheel and the work continuously. This can either be pumped, or held in a sump that the periphery of the wheel runs in and draws the water over itself.

The illustrations show the method of setting the angles on the tool rest of a simple bench grinder. There are much more sophisticated machines used by tool rooms with graduated slides and three-dimensional movement, but when grinding lathe tools infrequently it is more a matter of reproducing angles that have proved to work well than actually measuring precisely. Having found a suitable angle for the task, Fig. 2.14 shows a method of repeating it. What is more, light machining does not require a great variety of tools and, once you have set

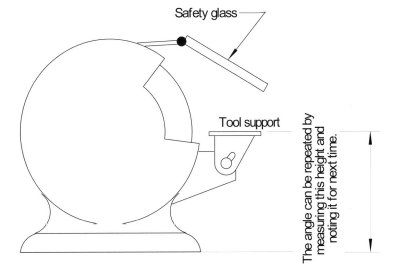

Fig. 2.14   Once you have found a suitable angle for a task, make sure you record it for future use.

Safety glass

Tool support

The angle can be repeated by measuring this height and noting it for next time.

the grindstone to produce cutting surfaces that work for you (and providing a standard HSS stick is used), you will rarely have any need to change the angle of the rest.

Grinding the tool will be easier and better if it can move smoothly and easily over the support. Make sure that the surface that the HSS tool sits on has no lumps and bumps; the example in Fig. 2.11 (although it was a good machine) had a cast support and it was not only rough but had a hard skin that a file could not deal with – it had to be ground.

Diamond Lap

This is a very useful finishing tool for HSS or tungsten carbide cutting tools and more use should be made of it. Diamonds (industrial diamonds) used to be very expensive and for this reason were only used where the cost of tools was less than the labour cost – in other words, where the value of the time saved was greater

that the tool cost. However, diamond files are much cheaper now and a simple hand lap or hone will cost about £15 (2020) and should last a clockmaker for a year or two.

Alternatively, diamond paste can be used on a grey cast iron 'carrier', which is relatively easy to file up and then polish with emery paper. The diamond for this comes in various grades of dust, already mixed into a paste with grease for applying to the carrier. It can be bought in small quantities (1g in a syringe) from clockmakers' supply companies from £5 to £10 depending on grade. The advantage of making your own carriers, or laps, is that they can be shaped to suit the job in hand. The cheapest commercial hand laps have a single thickness of diamond dust, deposited electrolytically, and a careless stroke that damages the cutting surface cannot be repaired.

Laps can be obtained from www.eternal-tools.com and www.mscdirect.com.

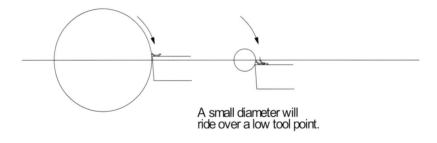

A small diameter will ride over a low tool point.

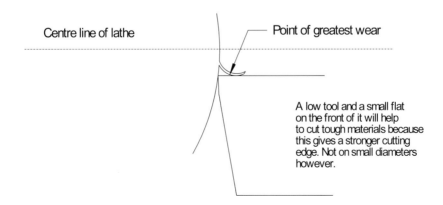

Centre line of lathe

Point of greatest wear

A low tool and a small flat on the front of it will help to cut tough materials because this gives a stronger cutting edge. Not on small diameters however.

Fig. 2.15   Always ensure that the tool point is positioned on a line that is radial to the centre of the rotating work.

TOOL POSTS

On the assumption that most people will make use of whatever tool post is provided until they have become dissatisfied with it, I have left the description of varieties of tool holding for the section on special attachments (*see* Chapter 6).

ALIGNMENT

The tool point should be positioned on a line that is radial to the centre of the rotating work. In the normal centre lathe, this can be achieved by ensuring that the height from the bed to the tip of the tool is exactly the same as the height of the headstock centre from the bed. If the tool tip is higher, it will only cut when well out from the centre. Large diameters can be turned when the tool point is a little high but small diameters cannot because the front face of the tool rubs on the turned diameter.

When the tool tip is lower than the centre, it will leave a small portion of the bar untouched after facing (a 'pip') and, when turning small diameters, a low positioning will cause the work to try to 'climb' over the tool. The latter effect becomes more pronounced as the turned diameter shrinks. For work of any given diameter, the error in alignment with the centre should be less than 5 per cent of the smallest diameter when turning.

Applying this guide, one should have the tip of the tool within 0.0127mm (0.0005in) of centre when turning a small pivot of 0.25mm (0.01in) diameter, or less than 0.127mm (0.005in) for a diameter of 2.5mm (0.1in). You will find from experience that this tolerance is tighter than absolutely necessary on larger work but quite vital on small pivots. When facing, the error should be kept to about 0.025 or 0.05mm (0.001 or 0.002in) – a 'pip' of this size will break off easily and not be a nuisance. Note that when the demand on the tool is great, the swarf or chips bear down on the top face of the tool behind the cutting edge.

Tungsten carbide is a sintered material and the work can pull out the hard material unless the metal cementing the carbide is calculated to resist this. Tungsten carbide tools are formulated for all sorts of duties: make sure that what you choose is the correct one for the task.

Iron and bronze castings that have a tough skin or sand inclusions may be roughed out to remove the hard layer by deliberately lowering the tip below centre. This will give a negative top rake to a normally ground tool. If the front clearance (relief) is diminished by grinding a short flat, the tool will not be greatly altered from its normal form and may be converted back to a regular turning tool with fair economy of tool material. Lock the cross slide while performing the roughing cut to prevent 'climb'.

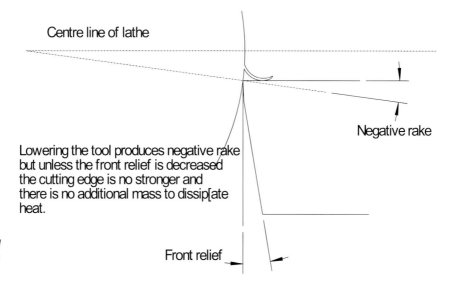

Fig. 2.16 Castings that have a tough skin can be roughed out to remove the hard layer by lowering the tip below the centre.

Centre line of lathe

Negative rake

Lowering the tool produces negative rake but unless the front relief is decreased the cutting edge is no stronger and there is no additional mass to dissip[ate heat.

Front relief

Grind a small 45-degree angle on the point of the tool also and approach the work at an angle – slowly.

SUPPORTING THE WORK

When the work (an arbor, for instance) protrudes from the chuck or collet by more than four or five times its diameter, it will often vibrate as the tool cuts and a poor surface is the result. Sometimes this can be corrected by taking small cuts and a fine traverse, but more frequently the work needs to be supported.

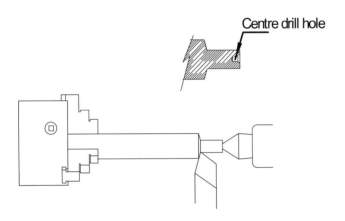

Centre drill hole

Fig. 2.17   Drilling a hole in the end of a bar with a combination drill bit creates a short cylindrical hole with a chamfer that should match the conical end of the lathe 'centre' that is usually held in the tailstock.

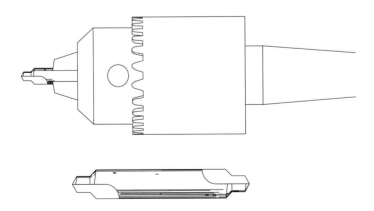

Fig. 2.18   Drill chuck and centre drill. These are held in the tailstock for drilling centre holes in work held by the three- or four-jawed chuck.

Fig. 2.17 illustrates one method. Here a hole is drilled in the end of the bar with a combination drill bit called a centre drill; it combines a short cylindrical hole with a chamfer that matches the conical end of a lathe 'centre' that is usually held in the tailstock. The drawing shows what a centre looks like; the part that fits inside the tailstock is normally a Morse taper, which is a standard taper developed in a range of sizes for lathes, drilling machines and millers. Fig. 2.18 shows a typical centre drill and a drill chuck with a Morse taper (not to scale) shank for fitting in the tailstock.

A simple centre is made in one piece from the Morse taper shank to the conical point and so it remains stationary while the work rotates. Friction between the stationary point and the rotating bar will damage the point unless it is lubricated and is pressed against the centre hole just hard enough to support without creating heat. However, there are centres available that have the point mounted on ball bearings, which allows the point to revolve with the work. The stationary centres are termed dead centres and those on bearings are called live centres and are much more convenient.

For the work that is carried out in clock repairing, this method is not possible and so an alternative is shown Fig. 2.19. Here a piece of brass rod has been drilled with a centre drill and used as a hollow centre to support the pivot directly.

Another means of supporting the work is called a 'steady', which is a very useful addition to the basic lathe.

When a lathe has been in use for several years, the bed will become worn in the area where the most sliding has taken place – usually close up to the chuck. At this point, the saddle will be loose and the tool will push away from the work centre, producing a taper. The only thing that can be said about this is that awareness of the problem will enable good work to be carried out still (particularly if a top slide is fitted) until the work demands that the saddle moves from the worn to the unworn position in the course of one job. Then the lathe should have its bed reground, or scraped with a hand scraper. Neither of these tasks is recommended for the inexperienced, but a skilled engineer who has served a full apprenticeship should be capable of doing it for a reasonable amount of money. Scraping takes longer than grinding, but is less likely to remove too much of the bed and it leaves a better working surface that retains any lubrication.

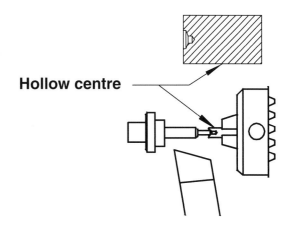

Fig. 2.19   *The upper image shows a piece of brass rod that has been drilled with a centre drill and used as a hollow centre to support the pivot directly. The lower one shows a rough collet positioned by a twist of wire and ready for soft soldering in place.*

Fig. 2.20   *A 'steady' for holding work on the lathe.*

Chapter 3

Facing, Boring and Drilling

The same tool that is used for turning may also be used for right-hand facing on small and mini machines. If there is a great deal of metal left on a face after cutting off the bar, the metal can be removed by repeated short lengths of turning before facing.

Then skew the tool to the left, move to the centre and make a light cut winding outwards to remove the steps left from the turning.

The relatively light facing that can be carried out with this turning tool will meet most requirements. A 'proper' facing tool cuts on its front edge. The cutting rakes and clearances are the same as for turning so long as the top rake is considered as an angle running back from the cutting edge either at right angles to it, or at an incline. The height of the

Facing with a turning tool making short strokes.

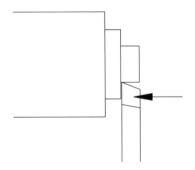

Fig. 3.01   Repeated short lengths of turning before facing will remove any excess metal.

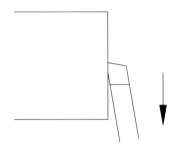

Fig. 3.02   Removing steps left from turning.

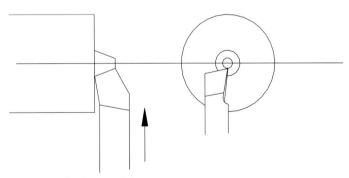

A proper facing tool for heavier work.

Fig. 3.03   A dedicated facing tool cuts on its front edge.

Fig. 3.04   Facing a wheel; the tool is being wound outwards.

cutting edge will become lower as the tool is reground.

Tool height is as important in facing as in turning: if the tool is too high, it will push off the face as it reaches the centre, and if it is low it will leave a pip.

Note: There are two designs of gauge for setting the tool on the centre line of the lathe shown in Chapter 7; both are easily made on the lathe and a good first turning task.

There is a strong tendency, when facing on a small or mini-lathe, to develop an undulation across the face (Fig. 3.05). This is caused by the traversing screw making use of the sliding clearance (and any bend or wear in the screw itself), to move the body of the slide from side to side as it travels across the face. The position of the cutting point in relation to the slide can exaggerate or minimize this.

If possible, move the tool point to a position where the least amount of movement results and, when finishing, use a 'shaving tool' approach to the work rather than a pointed tool (Fig. 3.06). This technique of shaving the work for finishing can also be used in turning, but simple honing will usually produce a good surface more rapidly (Fig. 3.07).

If the headstock centre line is not at right angles to the line of traverse for the cross slide, a concave or convex surface will be produced. A small amount of concavity (0.025mm/0.001in) or convexity is acceptable but any more needs to be corrected. This requirement is easily catered for on a lathe with a swivelling headstock, but if a small lathe is discovered to produce convex or concave faces, it indicates wear or looseness in the slide; consult somebody with machine tool experience with a view to correcting matters.

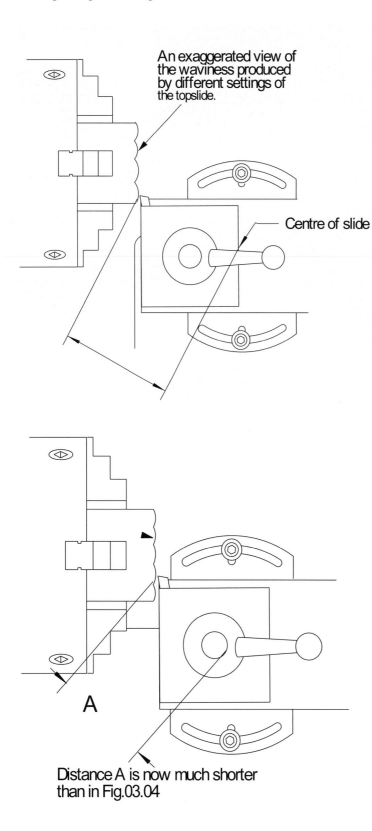

Fig. 3.05   Undulation across the face is a common problem.

An exaggerated view of the waviness produced by different settings of the topslide.

Centre of slide

Distance A is now much shorter than in Fig.03.04

A

Fig. 3.06   To avoid undulation, move the tool point to a position where the least amount of movement results.

Fig. 3.07 When finishing, use a 'shaving' approach to the work rather than a pointed tool.

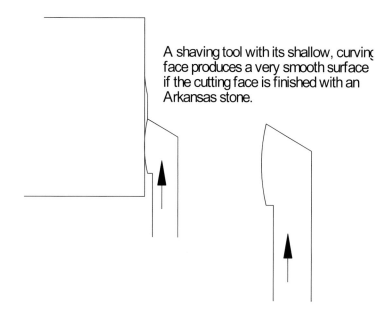

A shaving tool with its shallow, curving face produces a very smooth surface if the cutting face is finished with an Arkansas stone.

Alternative positions

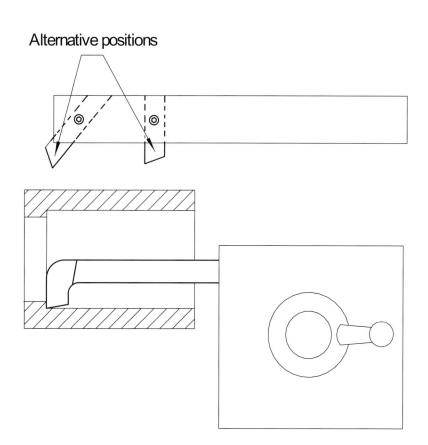

Fig. 3.08 When deep boring, the tool needs to be at least as long as the bore being produced. The cutting bit should be fitted into a bar that can enter the hole, or the cutting tool must have a long neck.

Fortunately, there are not many occasions when clockmaking or repair call for an accurately machined face larger than an arbor shoulder.

BORING

This operation may be divided in two – deep boring and shallow boring – with different forms of tool required for each. Unless the work is very large in diameter, only the actual cutting tool can enter the bore and consequently, when deep boring, the tool needs to be at least as long as the bore being produced (Fig. 3.08). This necessitates the cutting bit being fitted into a bar that can enter the hole or a cutting tool that is made with a long neck so that it can cut without the tool post touching the work.

In shallow boring no bar is used; in fact it is really a facing tool ground for machining from the centre of the face to the periphery. The tool is held directly in the tool post and protrudes from it by no more than the amount needed to machine a recess or counter bore. It makes no difference to the shape of the cutting face or the angles of relief or rake, but a shallow boring tool is more robust and less prone to 'dipping' (*see* below) when making a cut.

Making a Boring Bar
One of the frequent clock repair tasks is remounting a gear wheel, either because it has been damaged or because an error was made when mounting it in the first place. The hole in the wheel will be relatively small and a small boring bar is called for. In Fig. 3.10, a piece of rectangular metal (mild steel for preference) is being held in the tool post and drilled successively by a centre drill held in the three-jaw chuck and then a 2mm twist drill. The resulting hole will be on the centre line of the lathe and it should be about 40mm long.

The bar is now taken out of the tool post and held upright in a vice for sawing a slot parallel with the hole and just breaking through for half the hole's radius. The hole will have burr from the sawing and this needs to be removed by twizzling the drill by hand (hold it in a small collet called a pin vice).

This bar will be the holder for the boring tool, which is made by heating about 25mm of the end of a silver steel rod 2mm in diameter to red heat and bending it through 90 degrees. It is then allowed to cool down slowly by burying it in chalk or dry sand so that it does not harden. The end of the rod is shortened and filed to the form of a normal boring tool, then heated again to red heat, cooled rapidly in water (quenched) and the shank cleaned with emery paper.

To use the tool, slip it into the drilled and slotted body and put the assembly in the tool post, where it can be adjusted for protruding length and, by twisting it about its long axis, for the height of the cutting edge. Locking the tool is simply carried out by tightening the screws

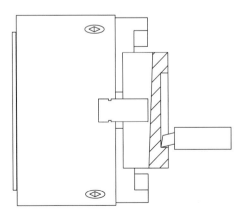

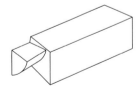

Simple, short boring tool for boring holes in relatively thin disks or for rebates.

Fig. 3.09   In shallow boring, the tool is held directly in the tool post.

Fig. 3.10 To create a small boring bar, a piece of rectangular metal is drilled successively by the centre drill held in the three-jaw chuck and then a 2mm twist drill.

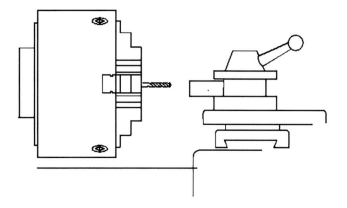

Drill a piece of rectangular section bar for a convenient silver steel rod (2mm dia.).

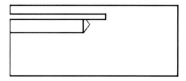

Cut a slot with a saw and insert the tool. It is held in place by tightening the toolpost screws.

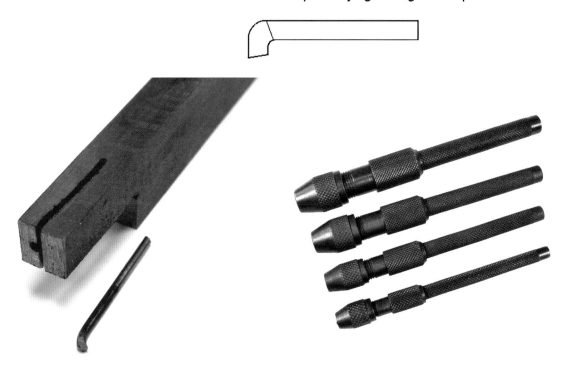

Fig. 3.11 A small hole boring bar.

Fig. 3.12 A selection of pin vices.

in the tool post. The sawn slot creates a spring that clamps the tool. After hardening the tool, stress relieve it by heating until the cutting faces display a very pale yellow.

Dipping

If a turning tool makes too deep a cut and tries to move away from the work, it will either jam or be pressed down (dip); in either case, the problem is easily remedied by loosening the tool in the tool post and removing it. Dipping in this case tends to relieve the pressure on the tool and the work.

However, when carrying out a boring operation, any dipping of the tool (because of the distance it protrudes from the tool post) makes matters worse if it has been aligned with the lathe centre. The tool digs deeper into the work. Placing the cutting edge slightly above centre obviates this problem.

DRILLING

There are several styles of small drill: spear point, twist drill, flat-sided drill and straight fluted drill. The first of these has no peripheral

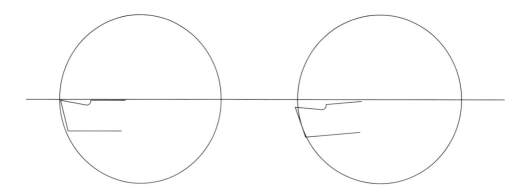

Fig. 3.13   Examples of dipping during boring.

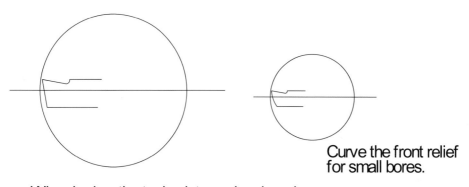

Curve the front relief
for small bores.

When boring, the tool point may be placed
above centre and will then relieve itself
if the cut is too deep.

Fig. 3.14   To avoid dipping during boring, place the cutting edge slightly above centre.

Fig. 3.15 Different styles of drill: spear point, flat-sided, fluted, flat or fluted, relieved and twist.

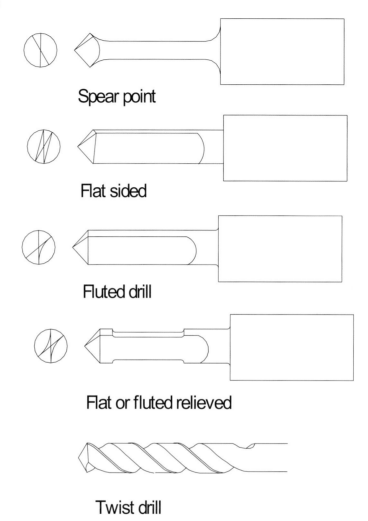

Spear point

Flat sided

Fluted drill

Flat or fluted relieved

Twist drill

surface to guide the drill and is best used for drilling holes in plate where a broach or reamer may be used later to 'size' the hole. Twist drills in very small sizes are rather fragile and, because they are made with a helix to evacuate swarf efficiently, they tend to lengthen slightly and dig in to the work if pressed a little too hard. The result is often a broken drill buried in an arbor.

Straight-sided drills are pointed like a spear point but they have a peripheral surface to hold the point in position by bearing on the hole being produced; the same applies to straight fluted drills but, because of the flutes, the drill has a greater cross-section and is stronger. These two types will work more easily if a relief is ground onto the outside surface about one diameter back from the start of the point.

Parting Off
Parting off can be carried out with a boring tool; Fig. 3.16 shows the method of cutting off thin walled rings or bushes after turning and boring. The same tool is used to make an inter-

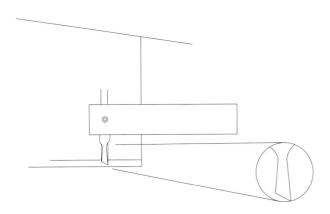

Fig. 3.16   Parting off can be carried out with a boring tool. Here thin-walled rings or bushes are being cut after turning and boring.

nal channel for a screw-cutting tool to run into when using a single point tool (*see* Chapter 5).

HOLDING SCREWS

Screws can often be gripped in the chuck with only a turn of 100gsm paper to protect the threads while the heads are turned or polished, but very small threads may be too small in diameter to grip. A short brass bar held in the chuck and then drilled and tapped to suit the screw can be used as a holding device, but if there is any appreciable amount of metal to be removed, the head must have flats filed on two sides or be screw-slotted first so that a grip can be taken on it to unscrew the work when finished. Use the original screw as a guide for deciding whether to use flats or a slot.

Chapter 4

Screw Threads

The previous chapters have dealt with the tools needed for simple turning, facing and boring, but even an inexpensive lathe is capable of a wider range of functions than this – and cutting screw threads is one of these. We should never underestimate a cheap or an old lathe; so long as a few criteria are met any lathe will perform accurately and produce satisfactory work. These criteria are:

- There should be no shake in the bearings of the headstock.
- There should be no movement of the tool due to excessive clearance or uneven surfaces in any of the slides.

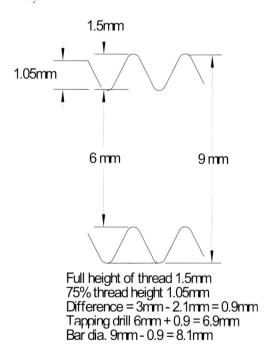

Full height of thread 1.5mm
75% thread height 1.05mm
Difference = 3mm - 2.1mm = 0.9mm
Tapping drill 6mm + 0.9 = 6.9mm
Bar dia. 9mm - 0.9 = 8.1mm

Fig. 4.01   The full (theoretical) thread for taps and dies.

- The stiffness of the headstock mandrel must be sufficient for the size of cut that it is intended to use as a maximum. It is useless to expect to reduce the diameter of a mild steel bar by 6mm (0.25in) at a pass of the tool if the diameter of the mandrel is less than about 25mm (1in).
- The tightness of the slides should be constant for the distance that the tool is to travel.

Fig. 4.01 illustrates the full thread for taps and dies. The tapping hole is chosen to leave a theoretical thread depth of 70 or 75 per cent of the full depth. In this case the full depth of the thread, or its height, is half the difference between the outside diameter and the hole diameter (minimum diameter): 9mm – 6mm = 3mm and half of that is 1.5mm. Seventy per cent of 1.5mm is 1.05mm. This is the height of the thread, which means that the prepared cylinder will be 6mm + 2.10mm = 8.1mm. The size of the drilled hole for tapping is increased by 2.1mm. Rod diameter is therefore 6.9mm and the hole for tapping would be 8.1mm. Since drills usually drill slightly oversize unless special precautions are taken, a 4.2mm drill would be satisfactory.

So far as clock- and model-making are concerned, there will be no noticeable lack of strength to a 70 per cent thread; even in more critical areas of use, the variation is little or nothing, depending on the amount of thread length in engagement. It helps to taper the mouth of the hole to ease in the tap.

Screw threads can be cut on the lathe in several ways: with die or taps, single-pointed tool, chasing – which depends upon developing hand skills – and hobbing and rolling. The last three are quite outside the scope of a clock- or model-maker's tasks unless they are proposing to manufacture in significant numbers.

The clockmaker will generally be using taps and dies.

TAPS AND DIES

In this case we are simply using the lathe as a useful means of holding the work and ensuring that the die or tap cuts square. Normally this will mean that the work will be held in the chuck and the tool either held, or at least supported, by the tailstock. However, this is not invariable, and there are many cases when it is easier to hold the tool in the chuck and put the work on the saddle.

When the tap or die is cutting, metal is often displaced to the top or bottom of the thread depending upon whether a tap or a die is being used, and it is common practice to produce a diameter of 70 or 75 per cent thread height to create a space for the metal to move into when the die or tap is used. The effect is greatest with metals that are not free-cutting and produce curly swarf, and least on metals that produce small chips (such as free-cutting brass). When tapping (producing a threaded hole), the drilled hole is larger than the theoretical dimension; when using a die, the turned cylinder will be small in diameter than the theoretical dimension.

Taps

Taps used to come in three shapes – taper, second and plug (or bottoming) – but most sets advertised in catalogues now consist of simply the second taps, or the second and a bottoming tap, which has a full thread right up to the

Fig. 4.02   *Cutting a screw on the lathe with the work held in the chuck.*

beginning of the thread. The holes that are most frequently tapped in clockmaking are 'through' holes in plates or holes in pillars, where it is easy to drill the hole much deeper than the threaded portion and a second tap will do the job.

Buy high-speed steel (HSS) ones with a ground thread as they keep their edge much longer than high-carbon steel and make a better job. The taps most likely to jam and break in the hole are those that have become worn and no longer have a good cutting edge. Using old taps is false economy. If the hole is 'blind' and does not pass clear through the metal, it may well be necessary to remove the tap after a few turns so that the swarf can be blown out. Shut your eyes when you blow! When the depth of the hole is limited by being blind and the thread is required to extend for most of its length, a bottoming tap will be needed after the second has been taken as far as it will go.

Die Cutting

Clock and watchmakers' die plates are still available and work very well on the smaller threads. Modern die plates often have the same sort of form as the split die that is used in a die stock – that is, two or three holes cut into the threaded portion in order to produce cutting faces, which, however, are not adjustable in the same way. Their main advantage is that one tool provides a number of different sizes of thread. A disadvantage is that the thickness of the die plates is tapered, and if it is supported by the face of the tailstock barrel, the thread will be skewed. Using a die plate on the lathe is entirely a hand operation, requiring the

operator to judge whether the plate is being held square to the turned bar or not. Split dies are preferable unless the plate is simply being used to clean up an existing thread.

Split dies (Fig. 4.04) are used in die stocks, and a two-handled holder that is a familiar workshop tool. The screwed portion is surrounded by three or four holes that cut into it and provide cutting surfaces, as mentioned above. The die is split in line with the centre screw of the three surrounding it so that it is possible to 'spring' the die to cut more loosely or more closely as

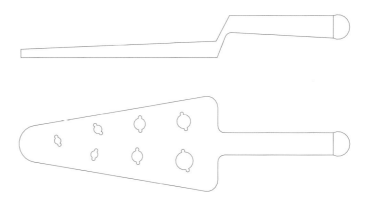

Fig. 4.03   An old-fashioned die plate, still used for very small threads.

Fig. 4.04   A split die kit.

desired. Where the cut breaks out of the outside diameter there is a V, and on either side of this two dimples.

The die stock has three screws: the centre one locates in the V and the other two in the dimples. These serve a dual purpose since, in addition to holding the die in place, they also open it up or close it. The screws in the dimples will close it up if they are screwed down hard, while the centre one will force the die open as it presses into the V. Obviously one set of screws has to be relaxed to allow the other to adjust the die size. Total adjustment is usually about three or four of the nominal diameter.

Nut Dies

These are really a millwright's tool and more use for correcting a damaged thread than in cutting a new one since they are normally held in a spanner, which, of course, does not guide the die very well. They are not adjustable and are usually made in carbon steel. However, they do turn up in sales and can be used if you are willing to make a hexagonal die stock to hold them. It usually means making a series of stocks, because the outside is a different size for each thread size, whereas a split die has a standard outside diameter that will suit a range of thread sizes.

Lubrication

Both tapping and dieing may need lubrication. Steels cut well with light oil, soluble oil (a water and oil mixture), and cutting pastes like Trefolex. Free-cutting steels, as you would expect, cut more easily than plain carbon steels. Free-cutting brass needs no lubricant, but 70/30 brass can be very 'luggy', which is relieved with paraffin or white spirit. Aluminium cuts better with paraffin.

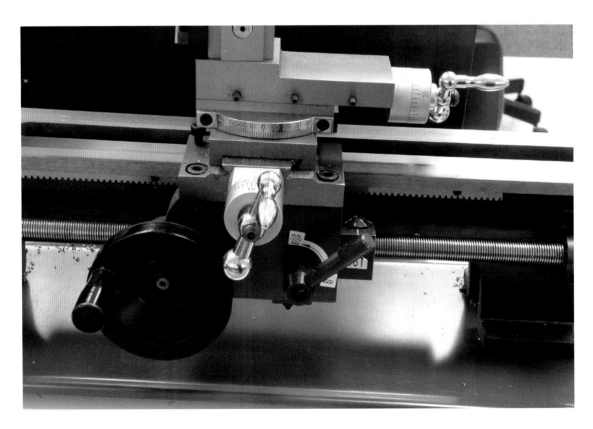

Fig. 4.05   A lathe's lead screw is the long screw that runs the length of the lathe bed and engages with a nut on the saddle.

Normal workshop practice with either tap or die is to advance the thread by half a turn and then turn back a quarter to break off the swarf that has collected, then forward another half turn. This is hardly ever necessary with free-cutting brass, but most necessary with any metal that tends to grip the tool, otherwise it becomes very easy to break a tap (in the case of threaded hole) or the bolt. Plenty of lubrication can relieve this, but the only way to measure the need to turn back is by judging the resistance to screwing. Acquiring good judgment can be an expensive business!

If the chuck is being used to drive a tap (with the work strapped down to the slide), it is better to use a four-jaw chuck and grip the square end loosely than to use a three-jaw gripping the round shank tightly. This allows the tap to move slightly and conform to the drilled hole. If it is held rigidly, any lack of alignment with the drilled hole will result in a broken tap.

SINGLE-POINT SCREW CUTTING

This type of operation is only possible if your lathe has either a lead screw or a feed bar, but for a workshop that has a large number of old clocks for repair, it can be very useful because many of the screw threads found in old clocks do not conform to modern thread systems. French clocks in particular did not make much use of the metric system until the twentieth century, instead using an old system based on the French 'inches' (ligne), which is close to imperial inches, but not close enough for the convenience of the clock-

maker. German clocks used metric threads in the nineteenth century.

A lathe's lead screw is simply a long screw that runs the length of the lathe bed and engages with a nut on the saddle. Fig. 4.05 shows this – it is the long bar seen behind the black hand-wheel; the nut (split nut) is out of sight behind the apron, which is the vertical casting that supports the mechanism for moving the saddle along the lathe.

A feed bar is a slotted rod that again runs the length of the bed but in this case drives a gear in the saddle by means of a key that slides along the slot. This gear in turn drives the traversing mechanism of the saddle. Mini-lathes often have a long slotted screw that serves both purposes. It is rarely necessary to use power for screw cutting in clock repair because the lengths of threads required are commonly no more than

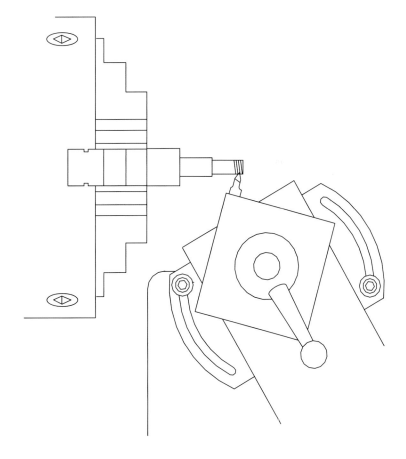

Fig. 4.06   Set-up of a lathe for screw cutting.

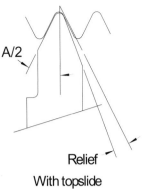 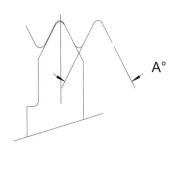

A/2

A°

Relief

With topslide   Without topslide

'A' is the angle of the thread

Fig. 4.07   The shape of screw-cutting tools for a top slide lathe and a non-top slide lathe.

two diameter of the screw. However, if your lathe has the facility for screw cutting, you will find a use for it sooner or later. The set-up of the machine is generally as in Fig. 4.06.

Fig. 4.07 shows the actual tool that produces the thread.

Choice of Gear Ratios

The mandrel of the headstock is made to drive the lead screw or the feed bar by means of a train of gears, so that a known number of turns is made by the lead screw for every turn of the chuck. Supposing the lead screw to have five threads per inch (tpi), every turn it makes will move the saddle 0.2in (5mm). If the screw makes one turn for one of the chuck, a single pointed tool will produce a complete thread every 0.2in (5 tpi). Increasing the number of turns that the chuck makes compared with those of the lead screw will produce a finer thread. For example, three turns of the chuck to one of the screw gives three threads for every 0.2in and, consequently, 15 tpi.

However, it is more useful to talk in terms of the count of the gear teeth on the mandrel (carrying the chuck) and the lead screw. If gears are attached to the lathe mandrel and the lead screw, we can write the following equation:

Tm = Number of teeth on gear attached to the lathe mandrel
Ts = Number of teeth of the gear on the lead screw

Pr = Required pitch
Pl = pitch of lead screw
Pitch is the distance between one turn of the thread and the next.

$$\frac{Tm}{Ts} = \frac{Pr}{Pl} \text{ and } \mathbf{Pr} = \frac{Tm \times Pl}{Ts} = \frac{60 \times 5}{20} = 15 \ tpi$$

If the train was a simple one, the ratio between them (1:3) could be obtained with one gear of 60t on the mandrel (chuck) and 20t on the end of the lead screw, and there would be one or two loosely mounted transfer gears meshing between them to bridge the space between the mandrel centre and the lead screw centre.

Transfer gears have no effect on ratios; an angular movement of a single tooth at one end of the train results in an angular movement of a single tooth at the other end of the train.

However, a finer thread than 15 tpi will require a larger ratio between chuck and lead-screw; for example, 1:9 would produce 45 tpi. If the smallest gear that will slip onto the end of the lead screw had 20t, it is extremely unlikely that a gear of 180t would be available for the mandrel. A simple train is then no longer possible and so compound gears are used (a compound gear is two gears fastened together on the same axis). The gear wheels B and C in Fig. 4.09 are the compound gears. Now we have a ratio of 1:3 from gears A and B and another 1:3 from C and D, giving 1:9 overall.

Fig. 4.08 Typical arrangement of gears to allow for reversing the rotation of the lead screw.

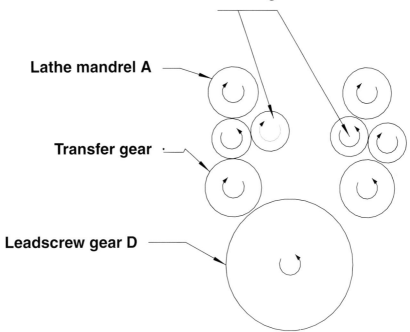

Tumblers for reversing train

Lathe mandrel A

Transfer gear

Leadscrew gear D

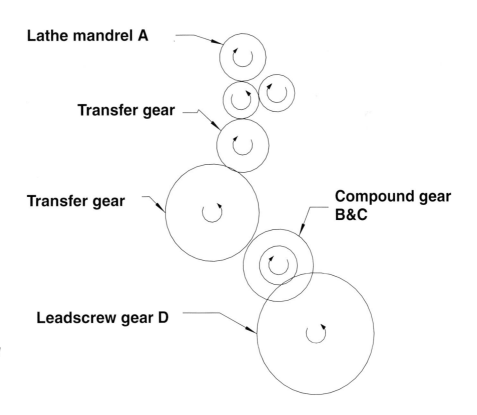

Lathe mandrel A

Transfer gear

Transfer gear

Compound gear B&C

Leadscrew gear D

Fig.4.09 Typical layout of the gears on the end of the headstock.

The smallest count of gears that are commonly available are 24 or 20. We can use the first with a count of 72 to obtain 1:3 and the train will then be A = 24; B, C = 72, 24 (compounded), D = 72.

The gears, A, B, C and D are called the change wheels. You will see that Fig. 4.08 also shows a simple means of reversing the rotation of the lead screw (the pair of gears E and F are called tumblers and they are mounted so that they can be swung into the train as a meshing pair or as a single gear). The lathe will also have a sub-plate that part of the train is mounted on and which is capable of being moved to alter centre distances. This is called a 'banjo' simply because it often had that shape (at least in the minds of deranged machinists).

Norton Gearbox

More expensive lathes avoid the need to make the calculations above and use change gears by arranging a selection of different counts on two parallel shafts that may be linked by moving a transfer gear horizontally and linking different gears. The outside of the box bears a plate with the screw threads available from the box and another lever engages what amounts to compounded gears to increase the number of screw threads available. All the operator has to do is to follow the instructions on the plate.

Metric Threads

Metric threads can only be cut if the lead screw has a metric pitch or a 127t gear is available. One inch is approximately 25.4mm (within 0.0001mm). A 25.4t wheel is impossible but 127t divided by 5 = 25.4 and this can be built into the ratio between headstock and lead screw. The metric screw threads are generally specified in terms of pitch, which is the reciprocal of tpc (or threads per centimetre).

SCREWING A SIMPLE BOLT

Turn the metal to the required outside diameter of the screw and then calculate the change wheels needed as described above. A thread that is machined close to the chuck may be made with no other support for the bar, but if there is any appreciable part of the work hanging out of the chuck, it must be supported either by means of the tailstock and a centre hole, or by a steady.

The included tool angles (see Fig. 4.07 again) for various thread systems are:

- BA: 47.5 degrees
- Whitworth and BSF: 55 degrees
- American, Unified and Metric: 60 degrees

Almost all threads since about 1950 have 60° flank angles.

A sharp point on the tool is not good. It should have a radius, or a flat (the American, Unified and Metric systems specify the proportions of

ABOVE: *Fig. 4.11   View of screw cutting in action and employing a steady.*

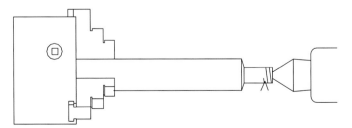

LEFT: *Fig. 4.10   Screw cutting a long shaft supported by dead centre (non-rotating centre).*

this) so that there is less danger of breaking the cutting edge of the tool and less of a stress concentration at the bottom of the cut thread. A radius of between 13 and 14 per cent of the pitch is common; screw-cutting gauges are available for judging the included angle, although they will not define the radius at the bottom of the thread. For that you will need a set of 'thread gauges' that show the pitch, the depth of the thread and the radius at its bottom. These come in sets and contain a fairly large number of gauges for different thread pitches. Another gauge style is needed to measure the angle of the cutting tool and the resulting thread; this is shown in Fig. 4.12b.

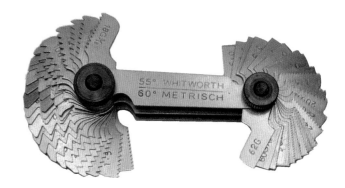

Fig. 4.12a   A set of thread gauges that show the pitch, depth of the thread and the radius at its bottom.

Clearances and Rakes

A screw-cutting tool that is to be held in the top slide should have side clearance and rake the same way as any other turning or boring tool (screw cutting can be performed inside a bore as well as outside a cylinder). Side clearances do not modify the angle of the V and consequently the flank angle of the thread, but top rake does. For this reason (unless the metal is easily machined and needs no rake at all) the tool is set in a top slide that is angled to match the flank angle of the thread, and the cut is applied using the top slide. In this way no rake is needed only top clearance.

In other words, the left-hand edge of the V-shaped tool lies on the lathe's centre line and this is the side of the tool that actually cuts. By advancing it at angle, the machinist produces one flank of the thread directly from the cutting edge, and the other by the point of the tool. The right-hand side of the tool may be allowed to merely scrape the flank (to smooth out traverse lines), or not touch it at all except at the start of the radius (which does not apply if the tool is flat pointed).

To ensure that the angle of the tool is correct, mount it in the tool post, making sure that the

Fig. 4.12b   A different gauge is needed to measure the angle of the cutting tool and the resulting thread.

point and cutting edge are level with the lathe centre. The gauge plate in Fig. 4.12b is used to check the position of the tool by laying one side of the gauge along the work and adjusting the tool until it matches the triangular cutout. Because it is a form that is being produced, any error in this height adjustment will affect the angle of the V cut in the metal – it is not simply a matter of reducing chatter.

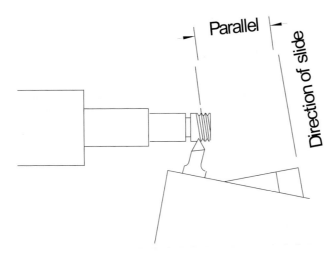

Fig. 4.13   The tool is set in a top slide that is angled to match the flank angle of the thread, and the cut is applied using the top slide.

Viewing the gauge from directly above will show whether the cutting edge contacts the gauging surface completely or not.

Note that if the lathe has no top slide, the tool has to be ground with a rake from the tip towards the tool post and will cut on both sides. Since the threads cut on a machine like this are most probably quite small in height (2.3mm/0.09in or less), the effect that rake has on the angle cut on the flanks of the thread is negligible.

Use a diamond lap to stroke around the radius (or the corners of the front flat) of the screw-cutting tool, not up and down, and make sure that you do not start to wear a groove in the lap or abrasive carrier (it is important to keep the surface that carries the paste dead flat so that the cutting edges of the screw-cutting tool do not get rounded over).

Machining Process Without Top Slide

We begin with the work already turned ready for the screw thread, held true in the chuck and supported by a running centre in the tailstock. The tool is held in the tool post so that it will traverse the whole length of the part of the bar that is to be screwed without fouling either the support or the chuck jaws. It should be set as already described for top slide machining above. This will ensure that the flank angles developed in the thread are symmetrical. The steps below assume that the lathe has no split nut and can only be recovered by winding the hand-wheel back.

1. Touch the surface of the work with the tool and make a note of the reading on the cross slide scale. Retract the tool and move to the right, clear of the work.
2. Advance the tool to the reading taken above and then increase this by the amount of 'cut' that you are intending to make. This will depend greatly on the material, strength of the tool and the diameter of the work, but for free-cutting brass of 0.5in (12mm) diameter and a pitch of 0.05in (1.2mm), let us use a cut of 0.015in (0.4mm) for this first cut.
3. Engage the drive between the lathe mandrel and lead screw; they will remain engaged until the thread is finished. It is at this point that you must decide whether to turn the chuck by hand or use the motor. For a small thread and a short run (less than two turns), I would prefer to turn by hand. It is a lot safer and the thread can be taken up to a dead stop. Some manufacturers provide a handle that engages the mandrel or the lead screw, making the job very simple.
4. Spread a little lubricant on the work and the point of the tool and then start turning the chuck so that the tool point gently begins its cut. Light oil lubricates brass and steel, but use mineral spirit for aluminium.
5. Continue turning the work until the tool has taken its cut as far as you wish the thread to extend. Note the reading on the cross slide and withdraw the tool, first allowing the chuck to turn back minutely. Otherwise there is a small risk that the point of the

tool could be broken off by the build-up of swarf.

6. Make sure that the point of the tool is well clear before rotating the chuck backwards so that the slide traverses back to the starting position and about to make contact. Take it slightly past this position, so that when you move forward there are one or two turns available to take up the backlash and bring the tool point to exactly the same position as in step 4.

7. Using the graduations on the cross slide, move the tool to the depth where it cut last time. Increase the cut by (in this case) about 0.01in (0.25mm). Rotate the chuck to make contact and cut the thread.

8. Turn the chuck as before until the tool reaches the previous stop. Withdraw the tool.

9. Repeat these steps, gradually lessening the cut as the point bites deeper until the required depth is achieved.

You will have noticed that the tool has been cutting on both flanks. This puts a load on it and there will have been a tendency to chatter, which is the reason for gradually decreasing the depth of cut. If all the cuts can be put onto one side of the V, it will cut better, with less crowding of the swarf, and leave a smoother finish on the thread. This is the reason for using a top slide when possible.

Relief Groove

There is very little time to halt the machine when screw cutting whether internally or externally, so cutting a groove for the tool to run into (Fig. 4.14) and rotating the chuck by hand or very slowly is well worth while. As an alternative (for an external thread), a hole the width of the thread is a neat method of proving a termination to the thread.

Finishing

The depth of the thread can be calculated from the relevant tables but, of course, the actual movement of the top slide was along an angle, so that it will be necessary to do a little trigonometry to discover what reading on the top slide scale represents the full depth. It is easier to make the screw to suit a nut (ring gauge).

The top of the threads will be flat. Round them off slightly with a smooth file when close to full depth, otherwise the sharp edge will prevent the nut from fitting even when the thread is fully developed. When the nut will just screw on (not necessarily all along the length of the newly cut thread) move the tool well away from the work, ensure that you cannot accidentally engage the lead screw nut and increase the speed of the headstock. The thread can now be polished with fine emery paper, backed by a triangular section of brass, until the nut fits smoothly. A grit size of about 400 should do for the emery paper.

Since you have not yet altered the train of gears connecting chuck and lead screw, it is possible to take a little more off with the tool if polishing will not make the nut fit. Do not break this train of gears between the headstock and the lead screw until you are certain the new screw is finished; it can be quite difficult to realign the tool when very little excess metal remains.

Internal Threads

An internal thread can be cut in the same manner as the external, but there are some important differences:

• The tool is mounted in a boring bar and will not be as rigid as the single-point or multi-point external tool. Take lighter cuts.

• You are usually cutting up to a stopping position that cannot be seen easily. Make a mark on the top of the boring bar to show the depth that you wish to go and finish each cut by hand, turning the work. Prepare the bore by cutting a groove or channel for the screw-cutting tool to run into. The channel should be at least as wide as one and a half pitches of the screw thread for the tool to run clear of the threaded portion. It would be wise to run the chuck very slowly or make the last turn before the 'run-out' by hand.

• Remember that to take the cut off the metal, the slide must be moved inwards. This is perfectly obvious intellectually, but as soon

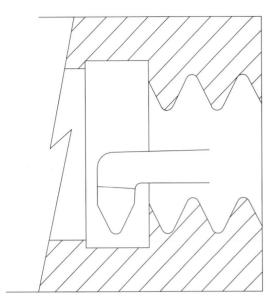

Fig. 4.14   *Cut a groove for the tool to run into and rotate the chuck by hand, or very slowly.*

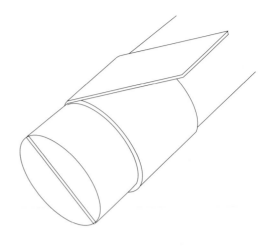

Fig. 4.15   *Remove the fraze with emery paper attached to a wooden dowel.*

as the excitement of approaching a blind end with a delicate screw-cutting tool begins to bite, you will be surprised how often instinct takes over and the hand screws the slide outwards – with disastrous results.

• Finishing the thread is a little more difficult, but it is just as necessary to remove the thrown-up fraze (burr) with emery paper as on the external thread. Fig. 4.15 shows a simple means of attaching emery paper to a wooden dowel.

Checking the internal thread is difficult. There are plug gauges that can be used to test the minimum or core diameter but they depend upon knowing (with some confidence) that the profile of the internal thread is correct. This is a difficult matter for a one-off job. It is simpler to make the male screw first and cut the internal (female thread) to suit it.

Both external and internal threads ought to start with a chamfer that is just a little deeper than the thread.

Chapter 5

Additional Machining Functions

TAPER TURNING

There are several ways of producing tapers on a lathe but the clockmaker rarely needs to make use of any but two methods, and the choice is governed by the type of machine. Mini-lathes frequently have a headstock that can be swivelled so that the axis of the chuck and the work are not parallel to the bed and the travel of the cutting tool; larger lathes are fitted with a top slide, which can be swivelled to perform the same function.

HORIZONTAL BORING

This is a technique that is of more use when making parts for tower clocks, musical boxes or

Fig. 5.01   Tapers may be machined on the larger machines by swivelling the top slide to the required angle and turning with a normal turning tool.

Fig. 5.02   Swivel the top slide to the required angle and turn with a normal turning tool to achieve a taper.

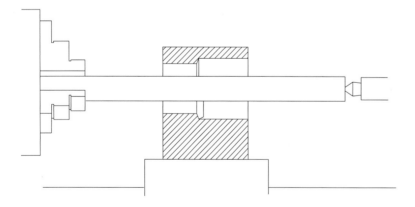

Fig. 5.03   A typical layout for horizontal boring. The work remains static and is mounted on the saddle or on a vertical slide.

models rather than domestic clocks. The work remains static and is mounted on the saddle or on a vertical slide. Fig. 5.03 shows a typical layout.

The only occasion that I needed to use this technique was in producing a 'bed' for the barrel of a musical box to rest on while I drilled for the pins.

A bar of steel is held in the chuck and, if necessary, supported by the tailstock centre. The cutting tool is fitted in a hole drilled through the bar and held with a grub screw. Behind the tool, the hole is tapped and a screw can be used to push the tool further out of the hole and thus cut a larger diameter. The work must have a hole in it first of all, so that the bar can pass through it. Since this sort of work is most often carried out on a casting, the hole is usually as cast. Moving the saddle traverses the work over the revolving tool, while movement of the cross slide or a vertical slide will shift the circular track of the tool sideways or vertically.

MILLING

This is a technique that is very useful if you intend to make tools and jigs for yourself, such as a Jacot tool, for instance.

End Mills

Held by the shank, an end mill can be used for cutting away a whole face, for milling a slot or a

rebate or producing the semicircular beds of the Jacot tool. Its speed must be kept down to the same sort of speed that a drill (HSS) of that diameter would be used at, about 30m/min (100ft/min) for mild steel. The usual method (needing no additional motor for driving the cutter) is to hold the end mill in the chuck or collet and the work either strapped down to the cross slide or mounted in a vertical slide.

Always ensure that the rotation of the mill does not cause it to 'climb' over the work. Only a very robust tool and machine is capable of withstanding this without a strong chance of breakage.

BORING HEAD

This is a chuck-held tool, or Morse taper-shanked tool, that has an adjustable cutting tool position. It is used for precision boring of holes or cavities. The example shown in Fig 5.06 has

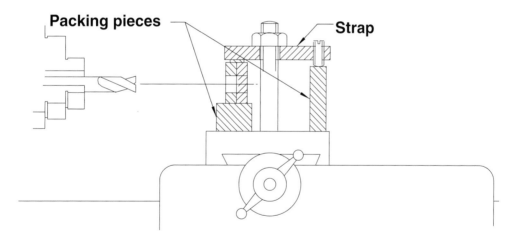

Fig. 5.04   Held by the shank, an end mill can be used for cutting away a whole face, for milling a slot or a rebate, or producing the semicircular beds of the Jacot tool.

Fig. 5.05   Make sure the mill cannot 'climb' over the work.

Fig. 5.06   A chuck-held tool or Morse taper-shanked tool that has an adjustable cutting tool position. RDGtools.co.uk

three forward-pointing positions for the boring tools that can be seen and one sideways-pointing hole for large-diameter work. Adjustment is possible to microns or thousandths of an inch.

FLY CUTTER

If this tool is mounted on the faceplate of the lathe, quite large surfaces can be faced square to the axis of the lathe. The size that can be faced depends upon the swing of the tool point, the total movement of the cross slide and the squareness of the cross slide to the headstock axis. A variation of the fly cutter is shown in Fig. 5.08, which is designed for cutting brass gear wheels.

The maximum height of the face produced is limited by the height that the rotating tool point can reach and, of course, this will also be affected by the way in which the work is mounted. The cross slide enables the tool (which is moving through a circular path) to cross the machined surface. It is not possible to machine a plane surface completely unless the traverse of the cross slide at least equals the diameter of the tool path. The largest area that can be machined is the traverse of the cross slide multiplied by the diameter of tool path; the path of the tool is of course circular, so the area will be a rectangle plus a half circle at each end. It is the rectangle that the above refers to.

If the headstock is not absolutely square to the cross slide, the surface cut will not be flat.

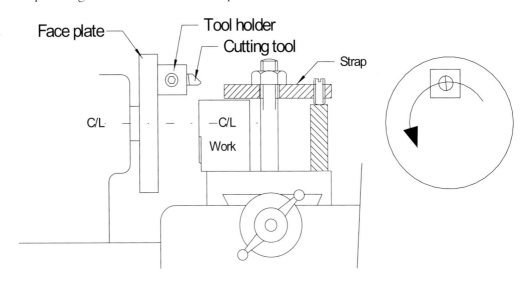

Fig. 5.07   A fly cutter.

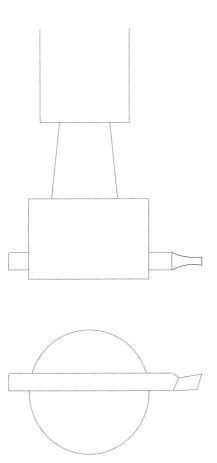

Fig. 5.08   A variation of the fly cutter that is designed for cutting brass gear wheels.

Fig. 5.09   Set-up of the lathe for gear cutting, with the work mounted in the chuck and the dividing mechanism attached to the mandrel. A separate electric motor or the headstock of a mini-lathe is used to rotate the cutter.

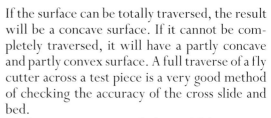

If the surface can be totally traversed, the result will be a concave surface. If it cannot be completely traversed, it will have a partly concave and partly convex surface. A full traverse of a fly cutter across a test piece is a very good method of checking the accuracy of the cross slide and bed.

Fly cutting is particularly useful for cutting gear wheels (*see* below).

GEAR CUTTING

There are several ways in which the lathe can be set up for gear cutting. The simplest for small gears is to mount the work in the chuck, attach the dividing mechanism to the mandrel and use a separate electric motor (or the headstock of a mini-lathe) for rotating the cutter. Although Fig 5.09 shows a directly powered cutter, there are many good methods of powering the cutter mandrel through a belt and remotely mounted motor. It really just comes down to convenience and financial expediency.

Holding the cutter in the lathe chuck allows the cutting of large wheels like those of a tower clock, but for domestic clock wheels the work is often obscured by parts of the lathe (Figs. 5.10 and 5.11). In fact, quite large wheels can be cut

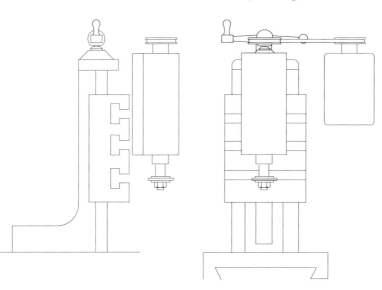

Sketch of vertical slide, cutter quill and motor, (Sherline headstock).

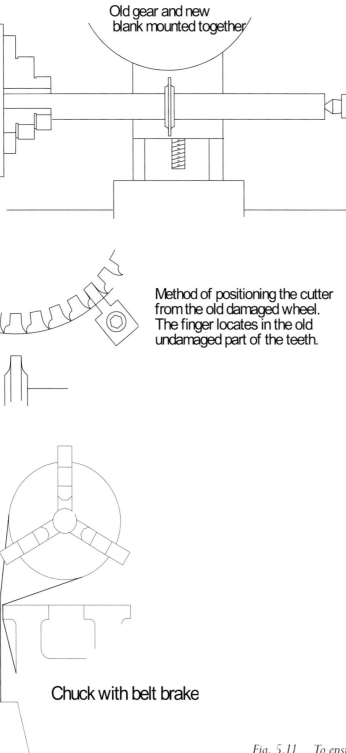

Old gear and new
blank mounted together

*Fig. 5.10   With the
cutter held in the lathe
chuck, large wheels like
those of a tower clock
can be cut.*

Method of positioning the cutter
from the old damaged wheel.
The finger locates in the old
undamaged part of the teeth.

Chuck with belt brake

*Fig. 5.11   To ensure that the chuck is held still
during cutting, a weighted strap can be hung over it.*

on the lathe by mounting the work on a vertical slide and carrying the tool in a boring bar supported by chuck and tailstock. The first gear that I ever cut for a clock was 15cm in diameter. I had the old wheel as a pattern and bolted this to the new blank. There were only two or three teeth missing and the bottom of the remaining ones was quite unworn, of course. I divided by simply locking into the bottom of the old teeth and got over the problem of the missing teeth by shifting the relative position of blank and old wheel halfway through the cutting process. The crossing-out took care of all the bolt holes.

Dividing heads can be bought or improvised.

Old gear wheels attached to the back of the mandrel can be positioned tooth by tooth, by using a pawl. The chuck should be held still during cutting, and a simple way of doing this is to hang a strap over the chuck with a weight on the end. This acts as a brake on movement and, when used with a dividing head, ensures that all the 'backlash' is taken up.

More flexibility in the 'counts' of wheel that can be cut may be obtained by mounting the old gear wheel on the lead screw mandrel and using the screw-cutting train to multiply the effect. For instance, an old 60t gear only allows the divisions of 60, 30, 20, 15, 12, 10, 6, 5, 4, 3 and 2. Using a train with a ratio of 12:1 increases this to all the factors that can be found in 720 ($2 \times 3 \times 4 \times 5$).

The dividing head is a most useful tool but it is expensive. However, reduction worm and wheel units can often be bought quite reasonably from engineer's supply warehouses, and

Fig. 5.12   Gear cutting with a multi-toothed cutter and a dividing head (left).

it is then only necessary to devise a means of supporting one of the standard dividing plates at the end of the worm and provide an indexing handle to obtain a very good dividing head. The most usual ratios are 30, 40, 50 and 60:1. The last has the largest number of factors in it, but I have managed with a 40:1 for many years and only had one awkward ratio to cut (63 teeth).

Fly-Cutting Gears

Multi-point cutters for gears are readily avail-able, but they are expensive. There is nothing that can be done about this as far as pinions are concerned, but wheels can be cut very satisfactorily using the fly cutter shown in Fig. 5.07. In fact, I prefer this method to multi-point tools. However, fly cutting only works well on relatively thin material, and for the clockmaker that means train wheels and great wheels but not pinions. Multi-point cutters are really needed to cut those.

Making a fly cutter tool is covered in Chapter 10.

Chapter 6

Chucks and Collets

Most turners will begin by making use of whatever holding device their lathe is fitted with. However, as time passes, you will find that there are limitations to whatever was originally provided, so let us discuss chucks, collets and faceplates.

CHUCKS

'Chuck' is a very general word but in metal turning it refers to a body of steel that carries holding jaws within itself. Other trades use the word to describe quite different devices.

Interestingly, 'chuck' in British English also means 'to throw', and an early hand-rotated version of the lathe was the clockmaker's 'throw'; and, of course, potters throw their clay onto what is in fact a vertical lathe.

The common three-jaw chuck is self-centring. This simply means that the movements of all three jaws are synchronized; any one of the three winding holes may be used to tighten or loosen the jaws They move in or out together and if the chuck is well made and in good condition, any piece of round bar that is held in the jaws will revolve 'truly' – that is, any turned portion will be concentric with the original cylinder. If a tool is lightly touched to the diameter it will scratch a complete circle around the bar with no breaks.

The standards of accuracy that one can expect from new three-jaw chucks larger than 75mm (3in) in diameter vary with price, but a standard workshop-quality 150mm (6in) chuck will run true and produce a total 'throw' of about 0.025mm (0.001in) within 25mm (1in) of the jaws, and 0.05–0.1mm (0.002–0.004in) at a distance of about 130mm (5in) from the jaws. A chuck of this size is not capable of securely supporting a bar that protrudes further than that.

'Throw' is the amount of eccentricity multiplied by two. If a cutting tool is set up to just touch the work at its furthest distance from the

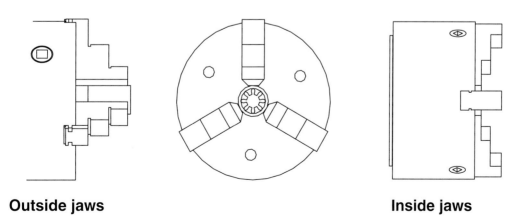

Outside jaws   **Inside jaws**

Fig. 6.01   The common three-jaw chuck is self-centring.

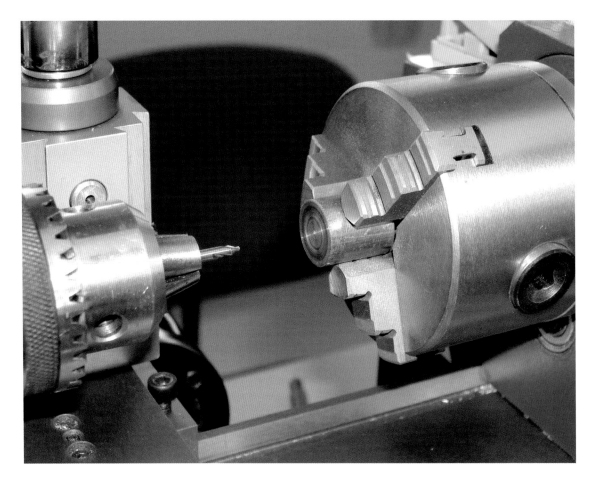

Fig. 6.02   A common use of a three-jaw chuck is to hold a work piece for drilling with a drill chuck (left) mounted in the tailstock.

lathe centre and then revolved by 180 degrees, the distance from the tool tip to the surface of the work is the throw.

Small chucks, as fitted to most mini-lathes, will not meet this accuracy unless they are modified by the owner.

It is normal to use a three-jaw chuck for holding material such as bar stock that has to be machined all over at one setting, or parts where the largest diameter is not required to be closely concentric with the turned parts. If a piece of work requires an outside diameter to be no more eccentric than 0.025mm (0.001in), the work must be machined all over without removal from the chuck; or some other arrangement must be made for holding it concentric to the previously machined areas when carrying out

subsequent operations. A small (75mm/3in) three-jaw chuck is not accurate enough to be able to machine both ends of a bar, for instance (requiring removal and re-chucking of the piece), and produce the two diameters concentrically. A 75mm chuck is very often 0.075mm (0.003in) out of true at a distance of 25mm (1in) from the jaws and not a lot better within that. This is because the type most commonly supplied with mini-lathes has jaws that are 'reversible', meaning they can be taken out of the body of the chuck, turned end on end and put back in again to grip on the outside of a larger diameter.

Since the jaws are moved in towards the centre by a scrolled plate, they carry a thread of square section on their backs. Large chucks have two sets of jaws, one interior holding and one

Fig. 6.03   Reversible jaws have threads on the back that are boat-shaped: they are formed from two sets of circles so that they can be turned around.

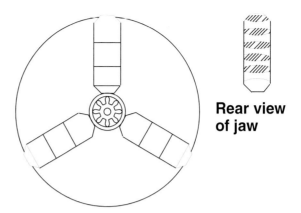

**Rear view
of jaw**

exterior, and the threads on the backs of the jaws are true parts of a circle (within limits), and will not allow the jaw to enter in more than one way. Each jaw is numbered and must be kept to the channel that bears that number. When in good condition, concentricity of 0.025mm (0.001in) is about the best that can be obtained.

Reversible jaws have threads on the back that are boat-shaped; they are formed from two sets of circles so that they can be turned around. When used in the reverse position, the jaw that was in channel 1 fits in channel 3 and vice versa; no. 2 stays in the same channel. Because of the excessive clearance brought about by the shape of the thread, the jaws are not controlled as completely by the chuck body as they should be

and there is a tendency for them to open up and become 'bell-mouthed'. Obviously, cylindrical work cannot be held true by a bell-mouthed chuck.

Correcting a Three-Jaw Chuck
Correcting a three-jaw chuck so that it runs true takes time and patience, but it is worth doing.

Materials needed:
• A piece of hard brass plate

Tools needed:
• Engineer's square
• Emery papers (200, 400, 800 grit)

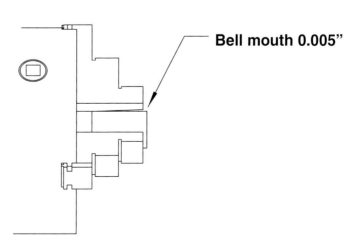

Bell mouth 0.005"

Fig. 6.04   There is a tendency for reversible jaws to open up and become 'bell-mouthed'.

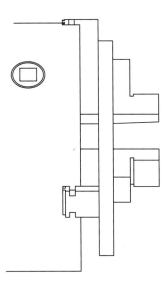

Fig. 6.05   For jaws that are to hold bars, it is recommended to immobilize them for correcting the gripping surface by making a ring to fit over one of the steps in the jaws.

- Emery or carborundum sharpening stone (medium coarse) or a flat metal surface to rest emery papers on
- Flat file
- Metal ring made to suit or a jubilee clip
- An independent grinding attachment (the motorized headstock of a mini-lathe strapped to the cross slide)

Fig. 6.06   A jubilee clip around the outside of the jaws will also serve to immobilize them.

There is a certain amount of play in a three-jaw chuck and the gripping surfaces have different alignment when they are pressing on the work from when they have been immobilized for grinding.

It is often recommended that jaws that are to hold bars should be immobilized by making a ring that will fit over one of the steps in the jaws and then opening the jaws and thus immobilizing them for grinding, or by using a jubilee clip around the outside of the jaws for the same purpose. This method is not exact and, depending upon the position of the ring, when the corrected chuck is used there may be a slight gap between the front or rear of the gripping surface and a cylindrical bar. For the correction to be effective, the jaws must be fixed in the same position that they assume when gripping work. The same is true for correcting internal jaws.

The following method is long-winded but more accurate.

1. Remove the jaws, and file up a piece of free-cutting brass approximately 3mm thick, that is about as wide as each jaw and with two faces that are at right angles to each other.
2. Support one of the jaws against the brass as shown in Fig. 6.07 so that when both are held on a carborundum stone, the back of the holding surface of the jaw bears on the stone while the belled mouth does not touch. Make sure the stone has a perfectly flat surface.
3. Stroke the jaw along the stone until enough metal has been removed for the whole length of the holding surface to bear on the stone – count the strokes.
4. Repeat with the other jaws.
5. Replace the jaws in the body of the chuck and, using a bright light, observe the closeness of contact made with a piece of truly cylindrical bar. If the bar shows light between itself and the front of the jaws, the guide ways of the body have worn and the brass guide must have the one face filed to tip the jaw and bear on the mouth. Use an engineer's square and feeler gauges to measure the amount of alteration that is made and go slowly – 0.05mm at a time. The process is time-consuming but, once the jaws come down onto the bar com-

Fig. 6.07   *Support the jaw against the brass so that when both are held on a carborundum stone, the back of the holding surface of the jaw bears on the stone. Put all the pressure on the jaw and periodically check the guide for wear.*

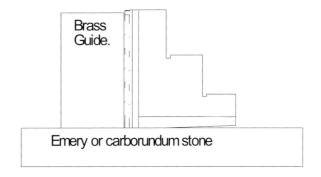

pletely parallel, there will be no need to alter the brass guide again.

6. The next step is stoning the individual jaws (using the brass jig) until a bar held in the chuck does not 'throw' further than a total of 0.025mm from the truly concentric. This can be measured with a dial gauge (Fig. 6.08) or by placing feeler gauges between the tool tip and a piece of undamaged silver steel. Move the tool in until it just allows the feeler to be withdrawn, then turn the chuck through 90 degrees and test for any change in the space between tool and bar. Repeat at 90-degree intervals until a complete revolution has been made. This test indicates how much stoning needs to be done. If only one jaw needs to be stoned, the amount to remove is half the eccentricity, but if two jaws are to be stoned, only roughly two-thirds of the throw needs to be removed. Do not attempt to remove it all in one go!

For the light work that is carried out on the lathes used in model- and clockmaking, there is no need to use a more robust three-jaw chuck than the normal scroll type. Heavier work is supported by a Taylor chuck, which has its scroll in the form of a shallow cone, giving a firmer

Fig. 6.08   *A dial gauge can be used to measure eccentricity.*

backing to the jaws and relieving the radial load on the scroll thread.

When work has to be held as true as possible to an already machined surface there are a number of methods that can be employed.

The Four-Jaw Chuck

These jaws are moved individually by separate screws. The jaws should be capable of tightening down on to a parallel surface and bear along their full length. They can be corrected in the same manner as the three-jaw version; however, there is no need to worry about eccentricity because the jaws are adjustable. The chuck is used on round bar, square bar, rectangular section, or any numbered section divisible by four.

Hold the work piece in the jaws and, using the concentric guidelines that are usually scribed on the face of the chuck, move the jaws in or out until the bar is held approximately on centre. Adjustment must now be made by rotating the chuck and checking the bar diameter or the corners against a pointer held in the tool post, moving one jaw at a time until a round bar does not show an increasing or decreasing gap as it rotates, or a rectangular bar touches the pointer at each corner evenly. If a jaw is being moved inwards, the opposing jaw must be loosened first.

When setting up a bar to an accuracy of less than about 0.075mm eccentricity, a pointer will be found to be too coarse an instrument. A dial gauge showing thousandths of an inch, or their metric equivalent, should be used and held to the bar whilst it is rotated by hand.

Other Types of Chuck

There is a chuck, usually called a combination chuck, that combines the advantages of four- and three-jaw chucks. It has a scroll and usually three jaws, but each of these is in two parts – the familiar stepped piece that grips the work and a sub-jaw that is moved by the scroll and which carries the other on a slide that is screw adjusted. The jaws can, therefore, be moved in and out by the usual key, but can also be adjusted for position individually; the individual adjustment remains unaffected by the scroll movement, so that this can be used to either obtain greater accuracy from a three-jaw chuck or to hold an oddly shaped or eccentric piece.

Yet another variation for both three- and four-jawed chucks is the soft-jawed chuck. Only the teeth and the guide slots are hard, while the part of the jaw that contacts the work is soft and bolted on. This is capable of being machined to suit the work but the scroll and the jaw teeth must be in good condition to avoid a bell mouth. They may, of course, be turned to fit inside an already machined bore. To maintain the accuracy, the jaws must be used on as limited a range of diameters as a spring chuck, otherwise the jaws will be registering on another part of the scroll and the original concentricity may be lost. Another

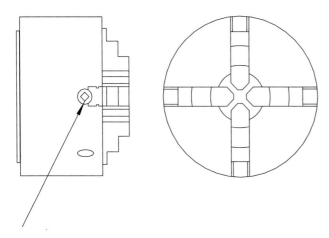

Fig. 6.09   The four-jaw chuck. Each jaw is moved individually by a separate screw (arrow).

point is that they cannot be adjusted very much after initial machining because, as they move in or out, the circumference that they are suited for changes.

COLLETS

As the name suggests, collets form a collar round an object to grip it tightly when a chuck alone is insufficient.

Ring Chucks or Bushes

A round-section bar that has had flats machined on it is often quite difficult to hold in a three-jaw chuck because the jaws do not register evenly on the bar – one at least falls into the flattened area. The ring chuck is a DIY version of the spring collet (see below).

Materials needed:
- Short length – 50–80mm – of free-cutting brass bar about one and a quarter times the diameter of the work that the collet is intended to hold
- A similar length of free-cutting brass twice the diameter of the work

Tools needed:
- Turning tool
- Boring tool
- Hacksaw
- Parting-off tool
- An internal scraper
- 800 grit emery paper
- Drill
- Flat file

Stage 1: the Ring
1. Reduce the diameter of the larger bar to twice that of the work that is to be held. Drill a hole 1mm smaller than the proposed outside diameter of the collet (*see* Stage 2 below).
2. Bore the hole to the diameter of the collet and bell-mouth the bore by about 1mm, using a triangular scraper (Fig. 6.10) You will probably have to make this yourself from a triangular file after grinding away the

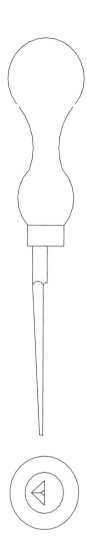

Fig. 6.10   Bore the hole to the diameter of the collet, and bell-mouth the bore by about 1mm.

teeth. Use it to enter the bore at an angle and adjust the angle until it cuts quietly – which it will do very quickly. Polish the bore with 800 grit emery paper.
3. Part the ring off the bar to be 3 or 4mm wide (Fig. 6.11).

Stage 2: the Collet
4. Hold the smaller bar in the lathe chuck with about half its length protruding; select one of the jaws and mark it with chalk or a felt pen where it will not be removed by the machining. (If the chuck has a label on its face, use that.) Dot the bar precisely opposite this mark so that if the bush has to be taken out of the chuck to saw a split in it,

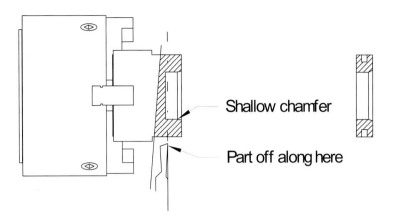

Fig. 6.11   Part the ring off the bar.

Shallow chamfer

Part off along here

it can be put back in exactly the original position. It also enables the collet be used again.

5. Turn the bar for at least half its length and reduce the diameter until when it is bored to fit the work, there will be a wall thickness of about 2–3mm.

6. If the work is long, drill right through the bar, producing a hole that is 1–2mm smaller than the diameter of the work. If the drilled hole goes right through the length of the bar,

the bored holed should do so too. The collet shown is intended for short work, such as a seven-leaved pinion.

7. Using a fine flat file to taper the outside of the bush slightly (0.1mm) until the ring will just slip onto it.

8. On the last cut through the bar, stop the traverse halfway along and, while the chuck is rotating, increase the cut a little – this is simply to make sure that long work does not bind up in the collet.

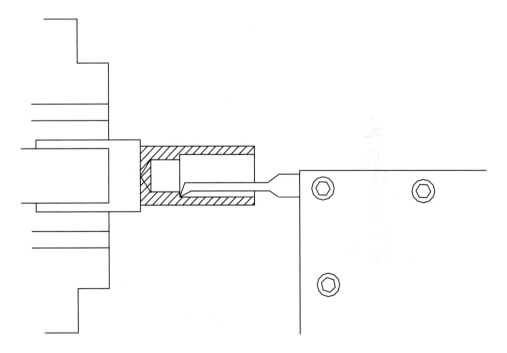

Fig. 6.12   Turn the bar for at least half its length and reduce the diameter so that, when it is bored to fit the work, it will have a wall thickness of about 2–3mm.

Fig. 6.13 To split the bar, fit the saw blade in the hacksaw frame so that it is a right angles to the normal position.

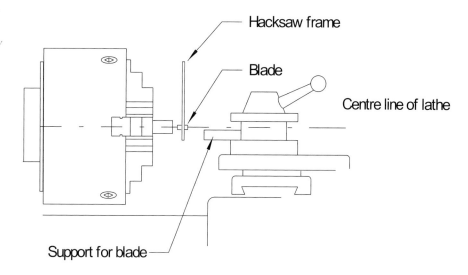

Hacksaw frame

Blade

Centre line of lathe

Support for blade

Fig. 6.14 The collet in use and holding a pinion head for drilling.

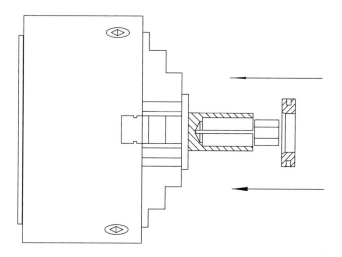

9. It must be split now. The best way to do this is to this to fit the saw blade in the hacksaw frame so that it is at right angles to the normal position. Fig. 6.14 shows the collet in use and holding a pinion head for drilling.

Spring Collets

A spring collet is a precision device that has a bore that is concentric with an outside taper. The taper engages with a prepared bore in the lathe spindle and the collet has slots cut along its length from the mouth. The slots are long enough to form springs and, when the collet is forced by a screwed cap or pulled by a screwed rod (draw bar) into the lathe mandrel, taper forces the sprung jaws to close a small amount and grip the work.

Most lathe manufacturers make arrangements for fitting spring collets (collet chucks) into the mandrel. Collets that are generally supplied for small lathes are commonly of the forms shown in Fig. 6.15 that allow the accurate holding of work (normally of round section) or tools such as milling cutters and drills. However, the method does not allow much movement of the collet jaws and so they are provided to suit nominal dimensions, or as blanks to be machined out by the purchaser. A collet of, for instance,

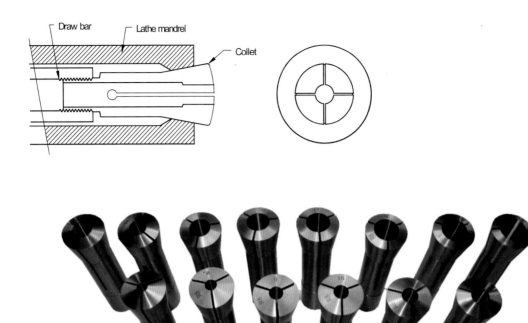

Fig. 6.15   Collets supplied for small lathes are generally of this form and allow for the accurate holding of work.

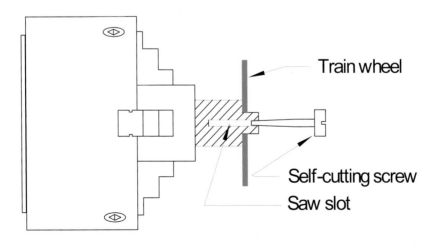

Fig. 6.16   An internal collet that is locked with a wood screw or a tapered self-tapping screw.

12.5mm (0.5in) diameter will only accept bar that varies from the nominal by about 0.025mm (0.001in), because the jaws are made as a piece with the rest of the body and, although they are springy, they cannot maintain parallelity beyond the specified dimension. These collets can be made to take all sections of bar and to hold on an outside or inside diameter.

Internal Collets

An internal collet that is locked with a wood screw or a tapered self-tapping screw is shown in Fig. 6.16. This is a useful method of holding gear wheels or any thin disks for trimming the outside diameter or polishing the face.

It is always difficult to machine an internal face to a dead centre, but if the depth of cut is accurately regulated with a stop, as mentioned before, or measured closely, a flat-ended drill can be used before the facing operation starts, to provide a cleared area in the centre for the boring tool point to start from, or the internal turning tool to finish at.

Rings and Washers

Internal parting-off can be carried out with a boring bar. It is particularly useful when cutting rings from a thin-walled tube (less than 1mm/0.04in thick). The boring bar serves to catch the parted-off ring with no risk of the work bouncing off the machine. It is also preferable to have the 'break-off' on the outside diameter if the ring has to slide over another part afterwards – parting with a boring bar achieves this (see Fig. 3.15). Note that a thin wall will crumple if too much load is put on the cutting face of the parting tool; to avoid this, the tool is ground to have a very pointed nose – 30 degrees or less. Fig. 3.15 shows a small flat at the tip which is no more than 0.2mm (0.008in) across; if the tip was very sharp, it would probably break off.

Chapter 7

Height-Centring Gauge

For effective turning, boring and facing on the lathe, it is important to know where the centre of the mandrel is so that the cutting edge can be placed properly. For most turning and facing, the machinist will set the cutting edge exactly on the mandrel centre. For boring, however, particularly for long bores, where the boring bar may dip when cutting, it is of advantage to set the cutting edge or point above centre so that if the tool does dip unexpectedly the depth of cut lessens and the load on the tool is relieved. If the tool edge was on the centre line of the mandrel – or worse, below it – the depth of cut would increase with dip and in all probability the cutting tool would be damaged. In the worst case, the work may be thrown out of the chuck.

A height gauge is a very useful accessory and

here are two designs that are useful and easily made:

FIRST GAUGE

This gauge is intended to define the centre line of the lathe measured from the lathe bed (Fig. 7.01). If the lathe has a gap under the chuck, machine the face of a piece of bar that reaches over the gap and use the smallest circular machining mark to determine the centre line.

Materials needed:
- A disk of free-cutting brass about 80mm in diameter and thick enough (20mm) to grip firmly in the three-jaw chuck

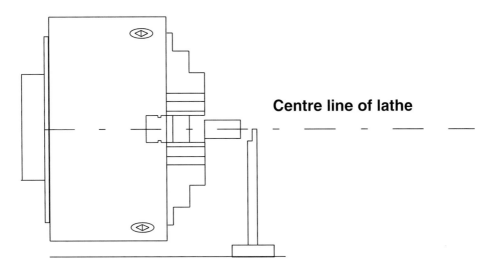

Centre line of lathe

Fig. 7.01   A height-centring gauge is intended to define the centre line of the lathe measured from the lathe bed.

Fig. 7.02 When making a height-centring gauge, grip the disk in a three-jaw chuck and pack three pieces of brass strip between the disk and the jaw faces.

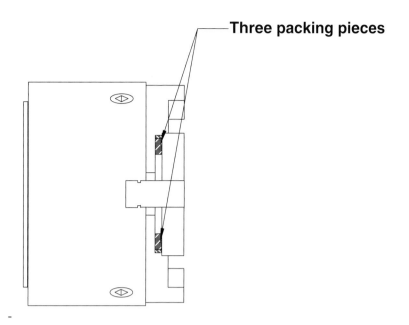

Three packing pieces

- A length of 5mm-diameter silver steel sufficient to reach from the bed of the lathe and slightly above the centre line of the lathe mandrel

Note: The chuck of a mini-lathe is not large enough to part off a disk from a rod 75mm in diameter. Possible solutions are to approach someone with a large lathe; purchase a disk sawn off a suitable bar; or buy a rectangular piece of brass plate 20mm thick (the base does not have to be round but it does need to have a rebate machined underneath). A disk sawn off a bar is much cheaper, particularly if you look to the future and order cut-offs that will be useful for jobs in the future. The set-up charge should remain the same.

Tools needed:
- Right-hand turning tool
- Sharp (preferably new or newly ground) 5mm-diameter twist drill
- 5mm-diameter centre drill
- Tap and tap wrench
- Die and die stock
- Loctite or similar adhesive
- 800 grit emery paper (plus brass backing strip)
- File

Stage 1: the Base

1. Grip the disk in a three-jaw chuck and place three pieces of brass strip cut from a bar (so that they all have the same thickness) between the disk and the jaw faces. This is to make the disk 'stand-off' the jaws and avoid damaging them with the tool (Fig. 7.02). A little thick grease will prevent the pieces from falling out while you press the disc hard back against them and the chuck. **Remove these packing pieces before switching on the power!**
2. Put a centre drill in the tailstock chuck and make a centre hole.
3. Change the drill for a twist drill that is the tapping size for the tap you have chosen and make a hole. Adjust the turning tool to make an angle of about 45 degrees to the face and pull it back, producing a cavity 2mm deep, until the original face is reduced to a ring about 3mm wide; this is to make the base stable in use (Fig. 7.03).
4. Use the tap to thread the disk.
5. Remove any burr on the circumference of the base.
6. Undo the chuck and turn the disk around, using the packing pieces as before. Make a chamfer purely for decoration and polish it

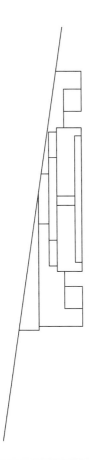

*Fig. 7.03
After making
a centre hole,
adjust the
turning tool to
make an angle
of 45 degrees
to the face and
pull it back
to produce a
cavity 2mm
deep. Ideally
the original
face needs to
be reduced to a
ring about 3mm
wide to make
the base stable
during use.*

with 800 grit emery paper backed up with a brass strip. To obtain crisp edges to your work when polishing with emery paper (cheaper and less likely to crack than emery cloth), always use a stiff strip of metal to back up the abrasive. Wooden emery stick are available and are useful but wood is soft, and it is easier to obtain crisp, well-defined edges with brass backing strips. The strips need to be hard rolled brass, 2 or 3mm thick and prepared with slightly radiused edges. From time to time, strike the edges against something hard to remove swarf and loose grits which may scar your work. Although the emery paper may be glued to the strip, I find it more convenient to wrap it end for end (doubled over the strip) and held in place by thumb and finger. When one side is worn out, it can be turned over and the strip used again instead of having to clean glue off and fitting clean emery paper.

Fig. 7.04   The split bush in use and holding a pinion in the three-jaw chuck.

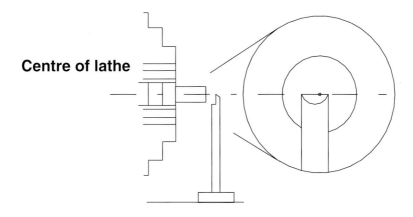

Centre of lathe

Fig. 7.05   Filing a flat on the upper end of the bar; make it a little longer than the centre line of the bar being held in the chuck.

Be fussy about the emery paper. Cheap paper often has badly graded grits so that (for instance) a 400 grit paper will have a few large grits, which will spoil the finish by leaving deeper grooves.

Stage 2: the Rod

7. Adjust the turning tool to face a piece of bar until the concentric rings are reduced to a point: this defines the lathe's centre.
8. Cut the 5mm rod to be a millimetre or two longer than the height; file one end flat and use the die to thread the other end. Before using the die, it should be adjusted to produce a thread that will be tight in the threaded disk. There are three screws in the periphery of the die-holder. Slacken the outer two and tighten the centre one; now cut the thread and test it in the tapped hole. When the male thread will enter the disk with difficulty it is just right. Make sure that the silver steel rod does not protrude more than about 0.5mm into the cavity.
9. Hold the bar in a vice horizontally and using non-ferrous vice guards, file a flat on the upper end of the bar and make it a little longer than the centre line of the bar being held in the chuck (Fig. 7.05). Use a file on the end to match it with the lathe centre; if it falls short, unscrew the bar from the base slightly. When you are certain that the centre is marked accurately, turn the tool over and put a drop of Loctite on the end of the screwed bar.

SECOND GAUGE

This gauge (Fig. 7.05) is designed for setting the height of a turning tool using a machined surface that is close to the tool post at the base; it needs to be relatively heavy and is best made from brass. In use, it slides over the point of the tool, which is adjusted in height until your fingers just detect contact as you move the gauge.

Materials needed:
- Short bar of brass 35mm diameter and long enough to stand about 10mm higher than the lathe centre when placed on the cross slide or the top slide (if fitted)

Tools needed:
- Right-hand turning tool
- Left-hand turning tool
- Vernier calliper
- Flat file
- Emery paper plus support

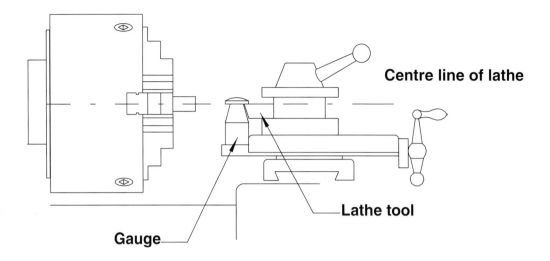

Fig. 7.06   The second type of height gauge.

Stage 1: Preparing the Rod

1. Use the right-hand turning tool and face both ends of the bar until the overall height is about 10mm more than the height from the supporting surface to the lathe centre (Fig. 7.07). Produce a 1mm (0.04in) chamfer on one end and a 3mm chamfer at the other. These are simply cosmetic operations but the face with the smallest chamfer will be the base of the gauge and should have a cavity similar to the one that was machined on the first gauge.

Stage 2: Checking the Height

2. Hold the end bearing the 1mm chamfer in the chuck with the end hard against the face of the chuck; this should result in the bar running true when the lathe is powered.

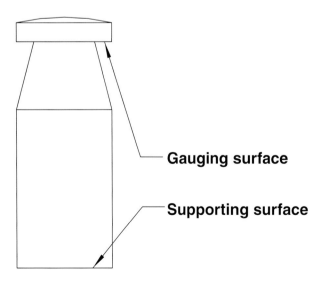

Fig. 7.07   Face both ends of the bar until the overall height is about 10mm more than the height from the supporting surface to the lathe centre.

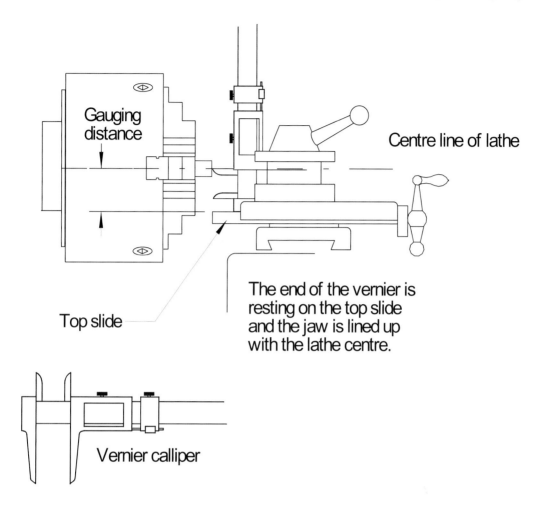

Gauging distance

Centre line of lathe

The end of the vernier is resting on the top slide and the jaw is lined up with the lathe centre.

Top slide

Vernier calliper

Fig. 7.08   Vernier calliper will give an accurate height measurement of the centre from the surface that you propose to seat the gauge upon when setting the lathe tools.

Check this and correct if necessary. Using the right-hand turning tool, take a very light cut across the face of the bar; this is simply to establish the true centre of the lathe.

3. Now use a vernier calliper to accurately measure the height of the centre from the surface that you propose to seat the gauge upon when setting the lathe tools (Fig. 7.08).

Stage 3: the Gauging Face

4. Change the tools and mount the left-hand tool in the tool post. This tool will rarely be used for producing a finely finished surface like that on a pivot so stone or grind a small radius on the point of the tool, being careful to maintain the front relief angle – a 1mm radius is sufficient.

5. A taper is now going to be made on the bar that will allow the left-hand turning tool to produce a gauging face (Fig. 7.09); this will also allow tools that are being locked in the tool post to touch the gauging face without touching any other part of the gauge and giving a false indication. As mentioned in Chapter 1, there are two common means of producing a taper on the lathe: a swivelling headstock on mini-lathes or a swivelling top slide on small lathes. The machining sequence is the same in both cases.

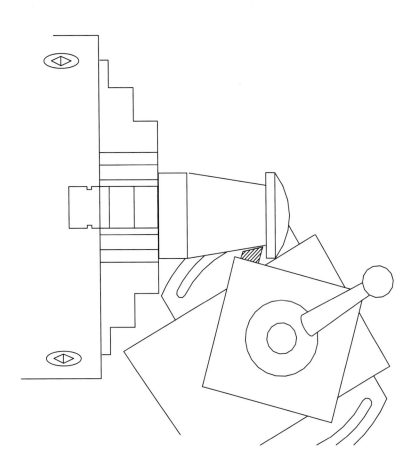

Fig. 7.09   Make a taper on the bar to allow the left-hand turning tool to produce a gauging face. This taper on the gauge may be used to check the angles of the tool's cutting side and front faces.

Make the first cut close to the gauging position and make sure that there is sufficient metal left at this point to allow for a facing cut when the taper is finished. Move the tool to the left until another cut of 1mm can be made. Repeat these cuts until sufficient space has been cleared to allow easy use of the gauge and provide plenty of surface to engage the sharp cutting edge without wearing away the metal.

6. The left-hand tool is now used to machine the gauging face; there is no reason to provide an undercut. Use a smooth flat file to remove sharp edges and then emery paper on a stiff brass or aluminium strip for support. Grit sizes will be successively 100, 400 and 800 grit. It will do no harm to protect against rust; wipe the gauge in any oil and then heat it with a gas torch until the oil smokes heavily. Leave it to bake on and, when cold, wipe away any excess with a rag.

Chapter 8

Pivots

Pivots, wheels and pinions are intimately linked and when any one of them is damaged, the others are often involved in the repair.

In the vast majority of clock movements, pivots are turned from the bar that the arbor is made from; some modern (twentieth-century) clocks have had the pivot made by crushing the end of the bar between rollers, which work-hardens the material, and a very few have inserted pivots of alloy steel. These are a boon for the repairer because a worn or damaged pivot may be extracted and easily replaced – presuming the repairer knows that it is an inserted pivot.

The final operation for well-made antique clocks was to heat-treat the turned pivot to harden it and then to polish it to a black polish. Mass-produced clocks were not hardened and either left with a fine-turned and polished finish or burnished. Polishing and burnishing are subjects discussed later in this chapter and are contentious depending upon whether the writer is most familiar with mass-produced clocks or British, French and German antiques with glass hard pivots.

NEW AND REFURBISHED PIVOTS

When a pivot (or its hole) is worn or damaged, there are several techniques needed depending on what has gone wrong; the materials and tools listed here (with, of course, the addition of the lathe) should cover all of these tasks.

Materials needed for complete arbor and pivots:
- Silver steel the same dimensions as the original arbor plus 2 or 3mm for machining to length

- Brass rod (machining quality) about 2mm larger in diameter than the hole in the wheel

Tools needed:
- Right-hand turning tool with a small (0.5mm) chamfer at the tip
- Right-hand finishing tool with a sharp point just smoothed with an Arkansas stone
- Emery papers 400 grit to 2,000 grit or crocus' paper
- Brass backing strip for emery paper
- Cutting broaches for 1–3mm holes
- 15cm (6in) flat file of smooth cut, fitted with a proper handle
- Drill chuck to fit tailstock
- Flat file
- Drills
- Vernier calliper
- Arkansas stone
- Pivot file

Pivot files do not have a rectangular section; the sides slope so that the cross-section of the file is a trapezium. Consequently, the face of the file can be brought right up to the pivot shoulder, cut a sharp corner and not rub the side of the tool against the shoulder (Fig. 8.01).

There is a difference in nomenclature between British turning practice and the Swiss makers of these files. The normal turned and faced surfaces produced by simple turning held in the chuck are termed 'right-hand' because they are on the right of the work piece. However, the file that is supplied as 'right-hand' or 'droit' will only file properly on a lathe if it is held under the pivot or with the tang away from the operator. I am not sure that all suppliers use the same nomenclature, so I would advise you either to tell them that a file is wanted to work on a pivot on the right-hand end of the arbor, or to visit the store and choose the file from the stock. Yet one would expect that the right-hand (droit) file was used rather rarely. Most lathes, or turns, have the chuck on the left and tailstock, or poppet, on the right. Flat burnishers for pivots are made and designated in the same fashion.

Pivot Faults

There are faults with a pivot that are not at all obvious and will need a good magnifying glass (or a sensitive thumbnail) to discover. Before I

Fig. 8.01   Cross-section of a pivot file.

Fig. 8.02   Another instrument for measuring heights or comparing one measurement with another was made using the lathe and hand tools from raw stock and a cheap micrometer.

Fig. 8.03   Different examples of wear to pivots.

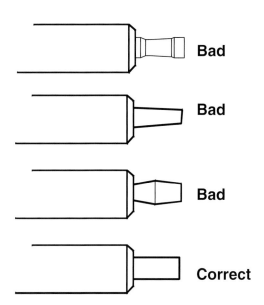

Bad

Bad

Bad

Correct

deal with replacing or even correcting a bad pivot, let us see what is required of good ones.

Firstly, they must provide low-friction support for the wheels and pinions.

Secondly, the support (that is, the pivot holes) must establish the proper engagement of the wheels and pinions, or the escape wheel and pallets or whatever else is borne on arbors or spindles. They must therefore be concentric when concentricity is required. (Pivots are sometimes carried in oval holes deliberately to obtain variation in some chiming work.)

Thirdly, they must be strong enough to do their job.

If the original pivot is broken or worn so badly (Fig. 8.03) that it has a visible 'waist', the question of what needs to be done is fairly simply answered: it must be replaced. Wear to this extent means that the pivot is weakened and that there is something embedded in the pivot hole that is cutting the pivot. Use a cutting broach to remove a small amount of brass from the inside of the hole and inspect with a [x]10 magnifying glass. If you have doubts about whether the intrusion is removed, you can expect that the hole will need bushing.

Refurbishing a Worn Pivot

On the assumption that the remaining pivot has enough metal to produce a good pivot, no material apart from the support is needed.

Note: if the pivot is a hard one, it will need to be annealed by heating to red heat and plunging into dry chalk or fine dry sand and leaving to cool very slowly for an hour or so.

Material needed:
- 6cm length of 15mm-diameter free-cutting brass

Stage 1: Support
1. The brass is needed to make a support for the pivots if one or both of them cannot be held so that the pivot being worked on is fairly close to the chuck or collet. Fig. 8.04 shows this support; it is simply a drilled hole that is a loose fit on the pivot that is then converted to a trough by filing away half the bar. It will be held in the tailstock chuck when being used.

Stage 2: Filing
2. Hold the arbor in the chuck or collet and lay the pivot in the trough (Fig. 8.05). Select a slow speed (about 100rpm); place the pivot file on the pivot so that it is trapped in the trough before switching on the lathe.
3. Stroke the file in the normal filing action and maintain light pressure on the pivot

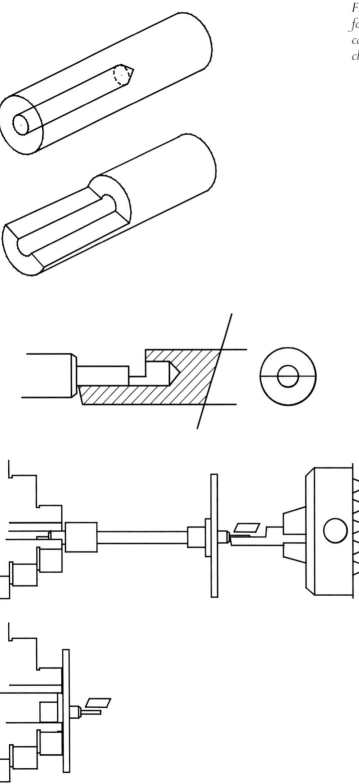

Fig. 8.04   A support is needed for the creation of a pivot if it cannot be worked close to the chuck or collet.

Fig. 8.05   Hold the arbor in the chuck or collet and lay the pivot in the trough. Select a slow speed and place the file on the pivot before the lathe is switched on.

Fig. 8.06 Keep filing until the pivot is completely cylindrical. It is not necessary to remove lines that lie outside the hole after shake is allowed for.

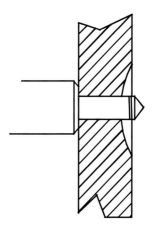

when making the back (retrieval) stroke to prevent the possibility of the pivot climbing out of the trough. Use the file until the pivot is truly cylindrical (Fig. 8.06). Any grooves that lie outside the pivot hole when the shake (longitudinal movement) has been taken up may be ignored – unless they are unsightly.

Stage 3: Preparing the Pivot File

4. One side of the pivot file is smooth when bought and needs attending to before it can be used. This is referred to as 'cut' and it is achieved by pressing the smooth side of the file on an emery board and sliding it from side to side as shown (Fig. 8.07). I prefer to use a grit count of 250, but use whatever count gives you the best result in terms of finish when used on the pivot – burnishing is greatly affected by the way the tool is used.

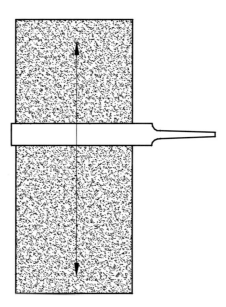

Fig. 8.07 Rub the pivot burnisher sideways across a board with 250 grit emery paper stretched over it to create the 'cut'.

Stage 4: Burnishing

5. Slide the tailstock away so that the trough and pivot can be cleaned of metal dust and grit, and then replace. The action of burnishing is different to that of filing. The tool is laid on the pivot and, at the same time as it is stroked forward, it is moved towards the shoulder while maintaining pressure (Fig. 8.08). Lubrication with a light oil or even saliva often works. Put simply, the aim is to smooth the surface by pushing the peaks into the valleys.

Turning a New Pivot

This exercise deals with the making of a complete arbor with pivots; for other situations, see Re-Pivoting and False Pivots below.

Materials needed:
- Silver steel or drill rod of the same diameter as the original arbor and 3 or 4mm longer overall.
- Diamantine (available from www.cousinsuk.com, www.hswalsh.com)

Tools needed:
- Gun metal or aluminium bronze strip 3mm thick × 50mm long and a little wider than the length of the pivot (for polishing)
- Flat file
- Vernier calliper
- Arkansas stone

Stage 1: Filing

1. Take the silver steel and remove any burr with a flat file; wipe it clean and either hold it in the three-jaw chuck or (preferably) in a spring collet. True silver steel is cut to length during manufacture by an abrasive wheel. The result is that a localized area at each end becomes hardened quite enough to damage the tip of your turning tool. Do not use the first 3 or 4mm of the full-length bar; saw it off and discard it.
2. Check that it is running true and, if it is not, take the collet out of the lathe mandrel and make sure that there is no dirt trapped in its seating. When you have checked that, do the same for the inside of the collet and the silver steel bar.
3. On the assumption that this arbor is 3mm in diameter, set the turning speed to 850rpm (or the highest speed that does not colour the tool tip or the swarf). If the length of the pivot to be turned is 4mm, have about 7mm sticking out of the collet. This is sufficient to avoid touching the collet or chuck and not so much that the work vibrates.
4. Use the file to make a shallow cone or con-

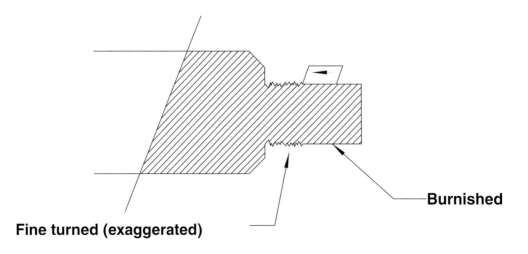

Fine turned (exaggerated)

Burnished

Fig. 8.08   To burnish, lay the tool on the pivot and at the same time as it is stroked forward, move it towards the shoulder while maintaining pressure. Lubricate with a light oil or even saliva.

vex surface on the end of the bar. When using a file on the lathe, you must make certain that if you slip, your hand (and the file) do not come in contact with any moving part. Regardless of whether you're right- or left-handed, hold the handle in the left hand and the end of the file in your right. Lean that arm on the top slide or the tool post or any convenient part that cannot move suddenly and make sure that you hand cannot move forward far enough to fall into the path of moving parts. Wrist movements are enough to make filing strokes.

Stage 2: Turning

5. Prepare one turning tool for normal turning, that is 10–15 degrees relief and a chamfer of about 1mm. Set it in the tool post with a slight inclination to the right

(Fig. 8.09). Tighten the locking screws and make sure that the underside of the tool is in full contact with the tool post with no air space (this could cause vibration or chatter).
6. Hold the bar in the chuck or collet with no more than 7mm protruding.
7. Switch on the power and use the handwheel to traverse the saddle along the bed smoothly and at a rate of feed that leaves a visibly smooth surface behind and with a depth of cut of about 0.75mm. The speed of rotation should be 800–900rpm. When traversing by hand, use both hands: one rotates the wheel or handle, while the other is ready to take over the motion when convenient. The idea is to maintain a steady rate of feed with no hesitation. If the movement stops and the tool is allowed to 'dwell' on

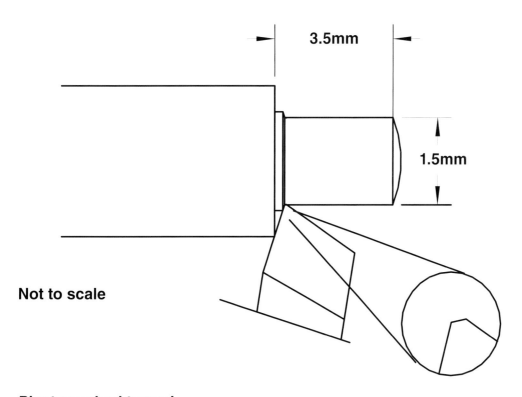

3.5mm

1.5mm

Not to scale

Pivot roughed turned

Fig. 8.09   Prepare a turning tool for normal turning. Set it in the tool post with a slight inclination to the right.

the work it will produce a line that is deeper than the other turning marks on the surface produced so far.

8. Use a vernier calliper to measure the length of rough pivot made so far (Fig. 8.10). The arbor is 3mm (0.1in) in diameter and the pivot is to be 1mm diameter and 3mm long. After this first rough cut, it should measure 1.5mm and be a little longer than 3mm.

9. The next cut is a finishing one and will reduce the pivot to 1.1mm diameter, which will leave 0.1mm to be removed by polishing. Any grooves that are left after this cut must be shallower than 0.05mm or polishing will not remove them, so, before making this last cut, take the tool out of the tool post and either exchange it for a finishing tool that you have already prepared or grind

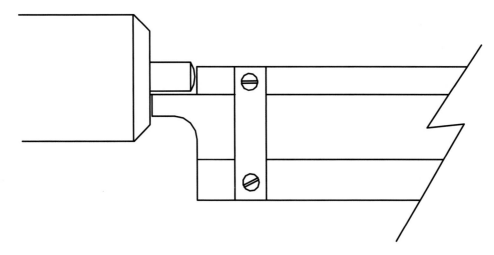

Fig. 8.10   *Using a vernier calliper to measure the length of the rough pivot.*

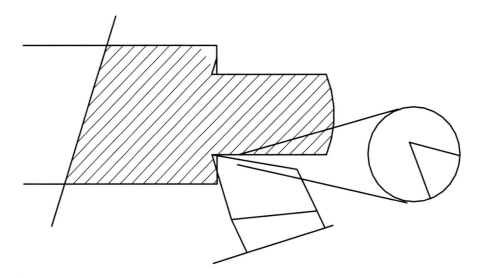

Not to scale

Fig. 8.11   *Do not allow the stone to rub over the tip and around the cutting edge.*

Fig. 8.12   Traverse the tool slowly to produce a cylindrical pivot to the correct measurements, with a spare 0.1mm on the diameter for polishing.

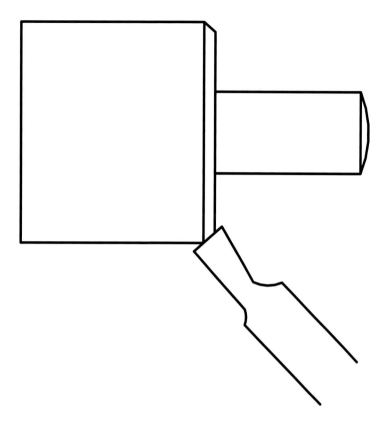

the front face back to eliminate the chamfer on the tip and produce a sharp point. Use the medium Arkansas stone on the upright edge between front face and side face for a few strokes and make a tiny chamfer of a hair's breadth (see Fig. 2.05). Do not allow the stone (sometimes called a slip) to rub over the tip and round the cutting edge (Fig. 8.11). Test the sharpness of the tip by dragging it across your thumbnail; its own weight should cause it to dig into the nail.

10. Replace this tool and tilt it to the left so that when the pivot is the correct length, it may be drawn out with the cross slide and produce a shoulder that is 'square' (truly at right angles to the diameter). Make a short cut at first to see if the tool is cutting correctly – the surface left on the cylindrical part should be smooth and shiny. If it is not, the tool needs to be re-sharpened and tested again. Remember that the silver steel rod is 2–3mm longer than necessary, so this test area can be filed away after final turning. Finish by making a slight undercut to

prevent the pivot corner from binding on the exit of the pivot hole.

11. Traverse the tool slowly to produce a cylindrical pivot to the correct measurements, but leaving 0.1mm on the diameter for polishing (Fig. 8.12). When this is finished, use a fine flat file or a parting tool to make a chamfer that leaves the face about twice the diameter of the cylinder (2mm).

On larger diameters of pivot, make the face diameter one and a half times the pivot. You need to balance the frictional effect of this face pressing on the inside of the clock plates; on the one hand, if it is too small, the pressure will be high and the shoulder will wear a shallow depression, increasing friction; and on the other hand, frictional torque will increase with the diameter of the shoulder. This is covered in more detail below.

Stage 3: Hardening
12. Silver steel or drill rod is capable of being hardened by heating to bright red heat (but

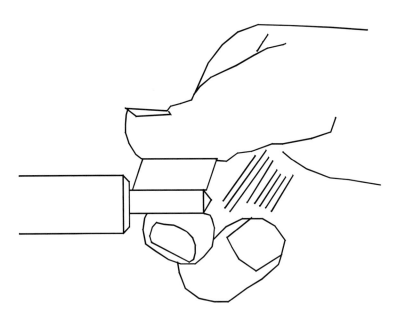

Fig. 8.13   Use a finger and thumb to squeeze the emery paper onto the pivot and slide it from end to end, keeping it on the pivot all the time and not rolling over the end.

Fig. 8.14   When finished, the pivot should bear a mirror finish.

not white) and cooling rapidly in water or oil. This will discolour the pivot but it is to be polished anyway so it does not need protecting. Cooling or 'quenching' in water produces a harder pivot than in oil, which cools more slowly. In either case, the metal is probably brittle and should be stress relieved after polishing by heating the middle of the finished arbor and watching the tempering colours creep along it until a very pale yellow reached the pivots – then quench in water.

Stage 4: Polishing

13. Use scissors to cut a strip of emery paper (or carborundum paper; there is very little difference in hardness of the grits) a little wider than the length of the pivot and cut a similar strip of brass that is no wider. If it is wider there will be a tendency for the combination of the two to roll over the end of the pivot and destroy its parallelity. Make sure that the chamfer has been touched with emery paper and will not cut your thumb.

14. Lay the strip on the pivot (the cut edge will go right into the corner) and use finger and thumb to squeeze the emery paper onto the pivot and slide it from end to end, keeping it on the pivot all the time – not rolling over the end (Fig. 8.13). Use successive grades of emery paper: 400, 800, 1,200 and 'crocus' paper (the term is recognized by stockists of engineering materials). When finished, the pivot should bear a mirror finish (Fig. 8.14).

Note: The pivot in Fig. 8.14 failed examination because although the finish was excellent, a small radius was left between shoulder and pivot – it should have been undercut.

The final polish can also be obtained with

diamantine, a white abrasive that is mixed with oil and crushed on a piece of flat and bright steel and a brass rod until it turns black; then it is applied with a prepared smooth strip of gunmetal or aluminium bronze and used in a similar manner to the emery papers.

To make the other pivot, take the bar out of the chuck or collet and paint it with engineer's marking ink, which will paint it blue. Set vernier callipers to the measurement between the clock plates when assembled without the 'works' and subtract 0.5mm for 'shake' end to end (the arbor has freedom to slide from side to side by 0.5mm). Locate one jaw of the calliper on the finished shoulder and use the other jaw to mark the position of the other shoulder.

Now repeat all four stages, using the mark as a guide, which must be checked as the new shoulder approaches the marking.

FRICTION

A very large amount of the friction that is developed in the movement of a clock depends on the design and finish of the pivots and pivot shoulders. From my own, rough practical investigations of friction at the shoulders of typical long and bracket clock arbors I attribute 50–60 per cent of the total frictional losses to the shoulders alone. Of the remaining sources of friction, teeth meshing and pivot bearings are the major components.

The pivots must have a polished surface at the shoulders and upon the diameters. For plated clocks, the cylindrical part ought to be a true cylinder, not tapered or barrelled (Fig. 8.15) because either of these faults can result in an unexpected binding of the pivot in the pivot hole – unexpected because it only occurs under certain conditions and most likely after the clock has been tested and returned to its owner!

As a result of the clearances that are normal for a clock, the arbors will be free to move backwards and forwards within the clock plates so that the part of the pivot that is actually carrying the load can change from one minute to the next. In fact, when a clock movement is assembled with no power on the train, all the arbors should slide from one side to the other as it is turned over.

If the pivots are not parallel (have the same diameter from one end to the other), then the meshing of the wheels and pinions will change as different parts of the pivot's diameter move in or out of the hole. In addition, a tapered pivot can increase the pressure against one shoulder and it increases the effective clearance between pivot and hole beyond what may be apparent from shaking the arbor. Clocks that have tapered pivots (posted clocks) have relatively large teeth that can accommodate the variation in meshing as the arbor slides from side to side.

RE-PIVOTING AND FALSE PIVOTS

Checking Pivots

First of all, you must decide whether the pivot as you see it now is the right diameter for the hole.

If you know that the clock has run for years and if the pivot hole is not oval and the pivot only needs light polishing, it almost certainly is the right size. However if it is rough and needs the use of a pivot file, this nearly always removes

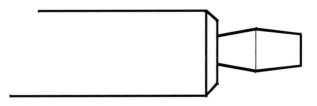

Fig. 8.15   For plate clocks, the cylindrical part ought to be a true cylinder.

Barreled

enough metal to make bushing the plates a necessity. As a rough rule of thumb, if the smallest diameter is about three-quarters of what you discover was the original size, the pivot needs replacing completely. The cross-section of a pivot that has been worn by this much is about half what it was and is very seriously weakened. If it is simply resurfaced it may bend or snap off in service.

A bent (and unworn) pivot can often be straightened by placing it in a hole that is an easy fit drilled in a piece of scrap plate and then gently bending it back into the vertical (Fig. 8.16). This is a chancy business in an antique, however, because the pivot may be very hard and brittle. Bending it back will probably be one trauma too many! Modern clock pivots bend more readily and if the degree of damage is slight and the bending has not taken place slowly during normal 'going' it is worth straightening the pivot. Bending as a result of normal working is a statement that the gear train is overloaded or the pivot is not fit for purpose.

If the pivot breaks off it must be replaced by one of the methods that follow. Re-pivoting arbors is rarely a pleasure. There are two ways of carrying out the job: drilling a hole in the arbor and inserting a piece of pivot steel, or making a false pivot for the end of the arbor that is then joined to the remains – I call this muffing.

Drilling for the Pivot

Materials needed:
- Brass rod one and a quarter times the pinion diameter
- Brass rod of arbor diameter (for pinions with seven or six leaves)
- Pivot wire (polished and blued)
- Silver steel of the same diameter as the arbor
- Loctite 601 or similar adhesive

Stage 1: Drilling
Inserting a new pivot calls for accurate and concentric drilling. The simplest method is illustrated in Fig. 8.17. The pinion being drilled is the centre pinion and it is being held with

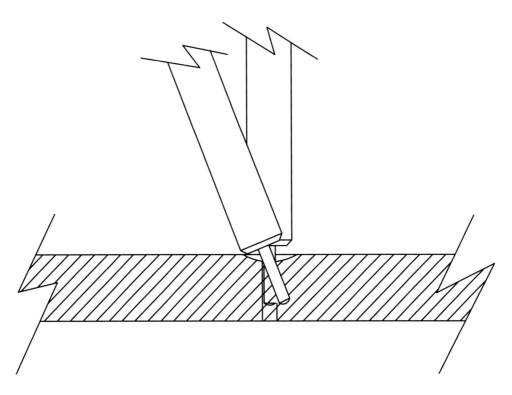

Fig. 8.16   Gently bending a bent pivot back towards the vertical.

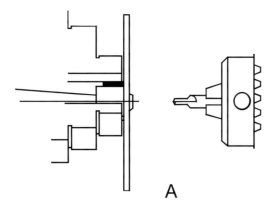

A

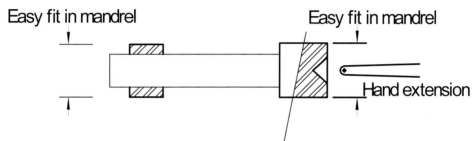

Easy fit in mandrel   Easy fit in mandrel

Hand extension

Repivoting a centre arbor; a split bush holds the pinion and
the other end is centred before the chuck is tightened by
a centreing bar which is a very easy fit in the lathe mandrel

B

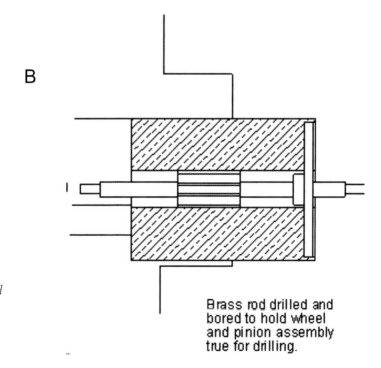

*Fig. 8.17   Two methods of
holding a pinion with mounted
wheel: A uses a split bush to hold
the pinion and a centring bar to
position the further end; B has
a wax chuck that ensures that
the wheel and drilled hole are
concentric.*

Brass rod drilled and
bored to hold wheel
and pinion assembly
true for drilling.

a temporary collet (a split bush). The end of the minute hand extension is centred before the chuck is tightened by a loose-fitting cent ing bar; this is simply a bar that either fits the bore of the lathe mandrel or has bushes fitted to make it fit loosely. Accuracy is not a problem: if it centres the extension to within 3mm, there will be no discernible effect at the pivot end.

Fig. 8.18 shows a method of ensuring that the drill enters the end of an arbor on dead centre. A piece of hard brass strip is held in the tool post and drilled with two diameters. The small one is the diameter of the pivot that will be inserted; the second one is the diameter of the arbor that is to be drilled and it only drills halfway. The result is two concentric holes: one to support the arbor and one to support the drill bit. Again, there is a rather more expensive tool available, a variation on the Jacot drum, that performs the same function just as well. The drills are held in the lathe chuck and support for the strip during drilling is provided by a tube

held in the tailstock chuck. The method is only useful for the end of an arbor that does not have a pinion or when there is sufficient length of arbor beyond the pinion.

Lubricate the hole in the support strip and set up the lathe as shown ; the small drill is now held in the tailstock chuck (Fig. 8. 19). Run the lathe at about 1,500rpm or its top speed and advance the drill, withdrawing it frequently to clear away the swarf. The depth of the hole must be at least two pivot diameters and preferably three.

Stage 2: Inserting the New Pivot

Degrease the hole and insert a drop of the adhesive before inserting the replacement pivot. The adhesive will go through a pasty stage for 5–10 minutes and if the pivot is steadied with your thumbnail while the lathe is spun, it can be made to run true and concentric with the axis of the arbor. When the pivot is running true, leave it spinning for a quarter of an hour and then take it out of the chuck. The Loctite will

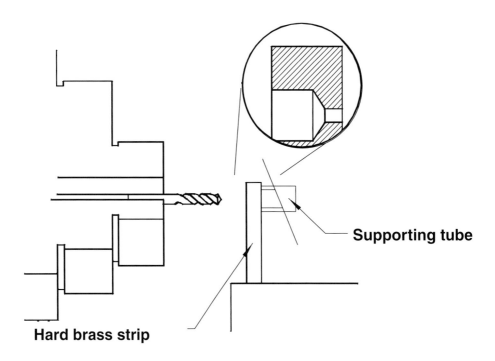

Supporting tube

Hard brass strip

Fig. 8.18   Drilling a brass strip with two diameters to support arbor and drill bit.

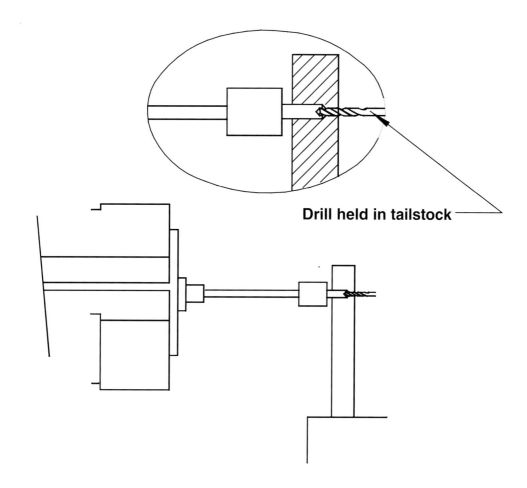

Drill held in tailstock

Fig. 8.19   Lubricate the hole in the support strip and set up the lathe with the small drill now held in the tailstock chuck.

develop full strength in twenty-four hours, but after an hour it will be strong enough for a saw-sharpening file or a mill file to be used on the sawn end and reduce it to the working length with a nicely domed end.

Annealing

Before attempting to drill, most antique arbors need to be annealed. Annealing is carried out by heating to red heat, holding for 10 or 20 seconds for a small pinion and then cooling as slowly as possible with normal workshop equipment. The problem is that some small pivots are extremely difficult to heat without doing damage to the fine leaves of the pinion. These are mainly from the high-quality steels used since the middle of the nineteenth century. Larger pinions from earlier clocks, such as longcase or bracket clocks, are made from steels that are rarely homogeneous. Because the steel was created by heating to bright red heat, folding and hammering, it has a very nasty tendency to contain hard spots, or sand inclusions that do not respond to softening. The result is that drills wander off centre and break. This happens to me about once in five times when dealing with eighteenth-century longcase clocks. When a wheel is fitted close to the end of the arbor to be annealed, it must be removed before heating (*see* below).

Muffs or Collars

Sometimes the arbor is so damaged that part of it needs to be replaced and the old pinion fitted to it. Muffing is a technique that allows the addition of a new length of arbor to what remains of the original (Fig. 8.20).

1. Mount the pinion and stub of arbor in a temporary collet (see below) and use a file to square off the broken end. Because the tem-

porary collet is probably not rigid enough to allow turning a reduced diameter on the stub, use a centre drill and a twist drill to make a hole two to three diameters deep.
2. Measure a length of silver steel to extend the drill arbor. Turn and polish a new pivot on one end and then machine a spigot on the other end to fit in the hole in the stub. When the two are fitted together, the measurement over the two pivot shoulders should be the distance between the clock plates minus

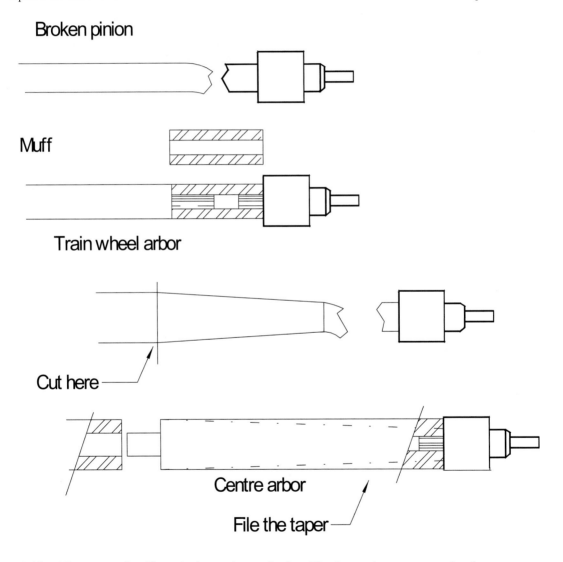

Fig. 8.20   *The process of muffing a broken or damaged arbor. The damaged part is removed and a new piece of rod is machined to fit onto or into a spigot formed on the pinion. If there is not sufficient length of pinion, it may be drilled to fit on a spigot turned on the muff. It is not possible to do this on a seven-leafed pinion because there is so little metal at the core of the pinion.*

Fig. 8.21 A hollow centre.

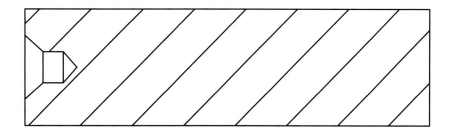

0.5mm (for shake). Leave the new arbor in the chuck.

3. Degrease and insert a drop of Loctite into the hole and slide it over the spigot, supporting it in a hollow centre held in the tailstock. A hollow centre is simply a piece of round brass bar with a centre hole drill accurately into the end (Fig. 8.21).

4. After allowing an hour for the joint to cure, spin the lathe and gently use a flat file to eliminate any mismatch where the two pieces meet, finally polishing with emery paper (800 grit). For safety's sake, continue to use the hollow centre.

Making a Temporary Collet Chuck

Place a short length of brass rod in the three-jaw chuck. It should be about one and a quarter times the diameter of the pinion's outside diameter. Face the brass, centre drill it and then drill out and bore until the pinion just slides in. Part off and remove any burr. The wall of this collet is now so thin that a little extra squeeze on the three-jaw chuck will grip the pinion firmly. Gently easing off the chuck jaws will release it again. This is a useful technique for many quick but fussy operations. Never throw a stub end of brass rod away!

Fig. 8.22 Take a piece of brass and turn the outside diameter to be 25 per cent larger than the pinion head that you wish to hold and for a length that is a little bit longer; part it off. Hold this piece in the chuck. Centre drill and bore until the pinion just slides in. The tube produced is so thin that if the pinion head is left inside it and the chuck is tightened, the metal will deform slightly and hold the head securely.

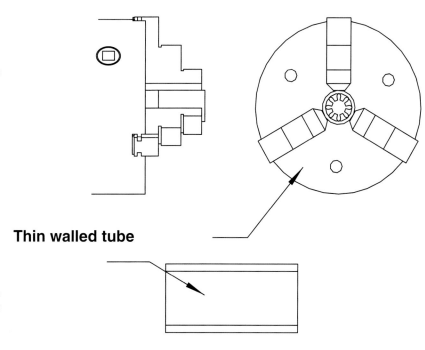

Thin walled tube

Pivot Ends

The conical end of the pivot is frequently used to run the arbor in a hollow centre or runner as, for instance, when running a pinion and wheel in the depthing tool. If there is no cone or if it is eccentric, the arbors cannot be set up properly to test the truth and mesh of the pinion/wheel pair. Old clocks often have pivots that are merely rounded with a file – persuading the arbor to run true in runners with this finish is difficult and sometimes impossible. A round end that has been turned and is therefore concentric will run true, but a cone is better, as it is not so adversely affected by slight damage to the end of the pivot. Do not neglect to harden and temper pinions and/or pivots after annealing and repair work. Dark amber tempering colour is needed for the pinion and it should be possible to cool before the pivot turns any darker than light amber.

False Pivots

When attempts to drill the end of an arbor for a new pivot go wrong or it has been damaged by bodging, the solution is to add a false pivot that fits over the end of the remains of the arbor and provides the pivot. The arbor has to be shortened to accommodate the addition and fit between the clock plates.

1. Remove the damaged end of the arbor using a saw or a file, and tidy up the end.
2. The false piece is bored at one end and turned at the other to form the pivot. Unless this is tackled in the right manner, the bore and the pivot will not be concentric and the whole point of making it will be lost. Fig. 8.23 shows the steel being drilled first to form the bore (A), and then a piece of brass being turned to produce a spigot that the false piece can be placed on (B) and locked in position with Loctite or similar. The spigot must not be moved in the chuck after it has been turned. Left as it is, it will be dead true, and the false piece that is pressed onto it will have both bore and pivot concentric. Move the spigot in the chuck for any reason and this will be lost.

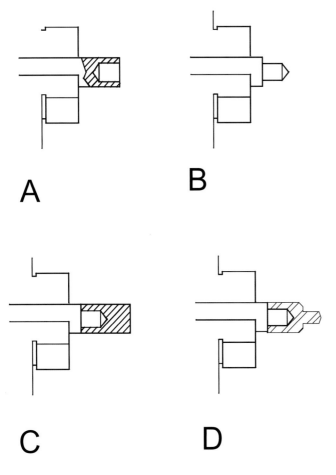

Fig. 8.23   A: a piece of silver steel (high-carbon steel) is drilled to accept a spigot turned on the end of the damaged arbor. It is then cut off to a length that will allow a new pivot to be made from it. B: the damaged arbor is shortened to allow for the addition of the false pivot and is turned to be a tight fit in the false pivot. C: file a slight flat on this spigot to allow air to escape when a cyano-acrylic adhesive is inserted and the new piece is pressed in. The spigot and hole should be at least one and a half times their diameter in length. D: when the adhesive has hardened turn the new pivot on the false piece.

3. Remove the finished false piece, take the spigot out of the chuck and set the arbor in its place (C). Turn the new pivot on the false piece (D).

4 If the wheel is attached to the end of the arbor that is to be held in the chuck, one of the accessories to the lathe is needed to support the arbor and wheel. This is called a steady and it is shown in Fig. 8.24. Adjust it until the pinion or wheel or simply the arbor is running on centre and then turn the damaged arbor until the false piece just fits over it. The length of the arbor will have to be altered to allow for the false piece.

There are two circumstances that call for special treatment:

• The end of the arbor is to be turned and the wheel held by the chuck because there is not enough arbor protruding at the wheel end to grip. A 'wax chuck' is needed.
• The arbor is not truly round or concentric with the pinion. A running bush is needed.

Wax Chucks

A wax chuck for gear wheels is simply a piece of metal that is bored with a very shallow hole or rebate that fits the outside diameter of a wheel.

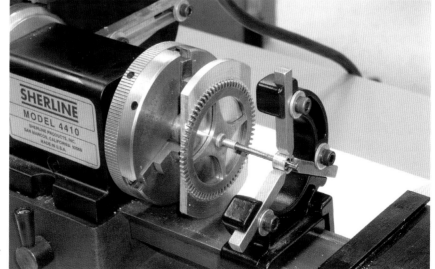

Fig. 8.24   A wax chuck for gear wheels is simply a piece of metal that is bored with a very shallow hole or rebate that fits the outside diameter of a wheel. Adhesion between the wheel and the chuck can be achieved using shellac or Loctite. It does not matter which; the only criteria to be satisfied are a plastic period to the setting, and easy release when a small amount of heat is applied. Above: the method when filing or burnishing the pivot. Below: for drilling a new pivot.

Adhesion between the wheel and the chuck used to be obtained by warming hard wax (sealing wax) or shellac, smearing it on the surface of the chuck and then pressing the wheel against it. Shellac is a good adhesive, as it remains plastic during its cooling and hardening quite long enough to make adjustments to whatever it is holding in place. You may prefer to use a modern material, however, such as Loctite. It does not matter which; the only criteria to be satisfied are a plastic period to the setting, and easy release when a small amount of heat is applied. Loctite 601 degrades to a powder at just over 150°C. Fig. 8.25 shows a wheel, arbor and pinion being held in the lathe so that the pivot may be attended to; the pinion is fitted into a running bush and supported by a steady.

Running Bushes

A running bush is a thin-walled cylinder that is fitted over a pinion (for instance) and locked there with an adhesive. It then rotates with the pinion, providing a surface to rest in the steady.

TAPERS

Before starting to file a pivot, use a micrometer to determine whether it is tapered or not. The faces of the micrometer anvils, being parallel to each other (Fig. 8.26), will show up any variation from a true cylinder when the pivot is lightly pinched between them and held up to the light. Take for example a pivot 1.25mm in diameter and about 3.5mm long; when tested

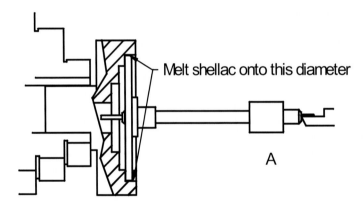

Melt shellac onto this diameter

A

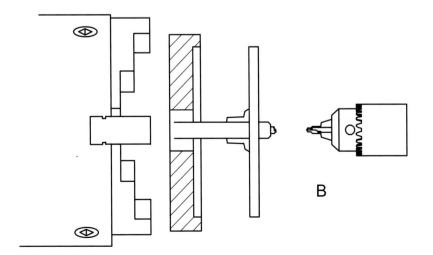

B

Fig. 8.25   A: the wheel of an assembled wheel and pinion is supported by a wax chuck, and a bush has been made by boring and turning at the same setting and has outside and inside diameters concentric. B: the steady has been closed on the bush, which will revolve between the three support fingers.

in this way, light will show clearly between the micrometer anvil and the surface of the pivot if a taper of only 0.05mm exists. All manner of pivot deformations will show up clearly in this fashion.

Such an amount of taper is not good, but it would not be so much that the repairer can justify reducing the diameter of the pivot simply to make it dead parallel. I cannot establish a provable rule for acceptable taper, but the amount just stated – 0.05mm in 3.5mm – is a very small included angle; it is reasonable to say that for most clocks, a taper like this will not lead to problems. After all, pivot problems arise from side pressure on the shoulders and changes in gear centre distances, both of these will be small enough to disregard. Twice as much as the example would have increased the centre distance (under the worst condition) by half the taper, 0.05mm.

When making a new pivot, it must always be a true cylinder unless intended for a 'birdcage' or posted movement. Those movement are dismantled and reassembled by locating the bottom of the posts in the baseplate, inserting one end of the arbor in a pivot hole and then swinging the post up until the second pivot can be inserted into its hole. The pivots are deliberately tapered to allow this. The pivot holes are tapered too, with the large diameter on the inside of the post (Fig. 8.28).

The likely tooth size or module of a wheel having a pivot of 1.25mm diameter is between 0.7 and 0.4mm. (Gear sizes or modules are commonly stated in millimetres in clockmaking.)

The amount of tolerance that we have on the centre distances of gears is about 10 per cent of the module. The taper of 0.05mm may increase the centre distance by half that (0.025mm),

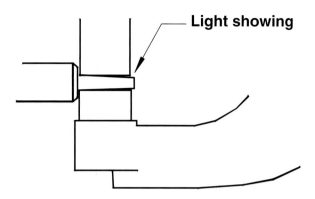

Fig. 8.26   Before starting to file the pivot, use a micrometer to determine whether it is tapered or not. The faces of the micrometer anvils are parallel to each other and will show up any variation from a true cylinder when the pivot is lightly pinched between them and held up to the light.

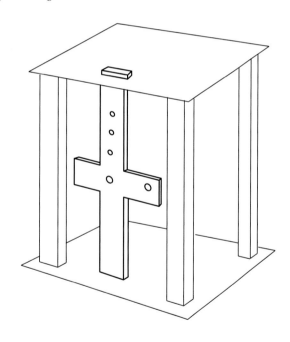

Fig. 8.27   Sketch of a posted or birdcage clock.

which is a considerable fraction of the 0.07mm tolerance that we have for centre distance of gears with teeth of 0.7mm module. It is added to the other tolerances on the wheel and the pinion and should be considered. When the arbor is in its worst position within the plates

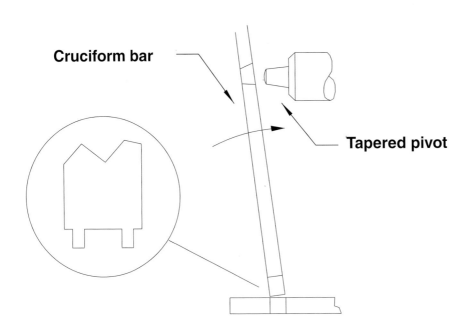

Cruciform bar

Tapered pivot

*Fig. 8.28
Reassembling
a posted or
birdcage clock.*

(with the small end in the hole), this taper has taken up most of the total amount of error that can be afforded on the wheel and pinion, meaning there is no tolerance left for any other error on the pivots of either of the gears that mesh with this pair.

RESURFACING THE PIVOT

Polishing

When the pivot is found to have a mere suspicion of roughness, it is quite acceptable to polish it with emery or crocus paper of very fine grit size: 2,500 grit will do nicely. Use the cut edge of the paper to get up to the shoulder and be aware that if more than 0.025mm is to be removed from the surface, it will leave a radius between the diameter and the shoulder. This amount of radius is acceptable, but it does mean that if you use emery paper and then discover that the diameter has been reduced by more than 0.1mm, you must use a pivot file to sharpen the corner again.

Pivots are often resurfaced while still in the hard condition; it is not necessary to anneal (soften) them first. 'Hard' as used here is a rela-

tive term – it does not mean dead hard as in British or European clock movements but hardened and tempered to blue or dark amber. The pivot file is normally quite capable of removing the small amount of metal needed to expose a clean surface once more without annealing the steel. The glass-hard pivots found in some high-quality antique clocks are an exception and it is very likely that they will need heat treatment first to soften or anneal them. There is a short piece on the heat treatment of pivots and pinions at the end of this chapter.

Filing

When polishing pivots with emery paper there is generally no need to support the work while it is being carried out because there is no reason why the operation should put much strain on the pivot. Using a file to remove metal is, however, a different matter and frequently does call for support.

The traditional tool that clockmakers and repairers use to support pivots is the Jacot drum (Fig. 8.29). This consists of two steel disks mounted one behind the other, on a fitting that attaches to the tailstock of the lathe. One disk carries a series of semicircular grooves

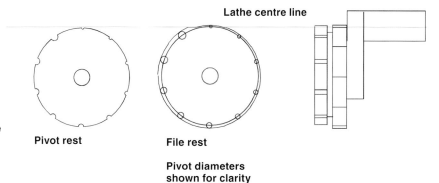

The pivot rest is mounted in front of the file rest and both are locked together. The pair can then be rotated to place the desired pivot diameter on the lathe centre line.

Pivot rest

File rest

Pivot diameters shown for clarity

Lathe centre line

Fig. 8.29　The Jacot drum is the traditional tool used to support pivots.

around its periphery, their pitch circle coinciding with the outside diameter of this disk. The tool is mounted so that the centre line of the lathe just touches the top of this pitch circle, so each groove is on lathe centre when it is brought to the top of the tool. Behind this is another disk, which is not a true circle but a series of arcs struck from the common centre with the first disk. Each arc coincides with the top of a circle that would represent the whole diameter of each semicircular groove and the diameter of the pivot resting within it.

To picture this more clearly, imagine a pivot lying in a groove. The groove is chosen to be of the same radius as the pivot (or slightly larger) so that the pivot lies with its lower half sunk in

the first disk; the second disk is so positioned that if a file is rested on the pivot and the disk at the same time, the surfaces touched are the same distance from the disks' common centre. Thus the file will cut parallel to the centre line of the pivot and arbor. It is relatively easy to file a pivot properly in this fashion if the second, supporting disk can be rotated to bring up a surface that compensates for any difference between the pivot radius and the groove used. If a large amount of metal has to be removed because of heavy wear, the groove must be changed to allow for the new radius of the pivot. However, that situation is most likely to require re-pivoting.

The type of support shown earlier (Fig. 8.03)

With Jacot drum

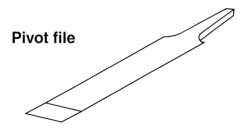

Pivot file

Fig. 8.30 Section through a Jacot drum to illustrate the difference between the two disks. The difference between gauche and droit files is shown as well.

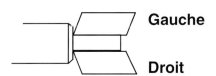

Gauche

Droit

Fig. 8.31   Using a pivot file on the lathe to smooth the shoulder of the pivot. The file is upright and the cutting face is working on the shoulder's face.

does not prevent a tapered pivot from being produced, but it is easy to make and very little practice will result in parallel filing without the need for a secondary support.

Fig. 8.32 shows three ways in which the pivot may be presented for filing. Sometimes a firm grip by the thumb and forefinger (as was shown in Fig. 8.13) is sufficient unless the pivot is finer than about 0.75mm in diameter (Fig. 8.33). Frequently, though, there will be very little that can be held firmly in chuck jaws, and then the Jacot or the alternative is the only possible way to proceed safely. This applies particularly to antiques, which often have arbors that are no longer true cylinders but barrelled or tapered. It is not merely a matter of making sure that the pivot is not bent or broken by the action of filing, but of ensuring that the arbor does not fly out of the machine.

If the arbor is held at the extremity of the opposite end to that being filed, you must use a support, and if you do not trap the pivot under the file before starting the lathe you will regret it. Inside a very few revolutions the pivot will climb out of its little bed and the arbor is very likely to be bent – you may even damage the wheel and pinion.

After taking a first cut over the pivot with the file, examine the surface carefully. This stage of refurbishing the pivot should continue until no grooves

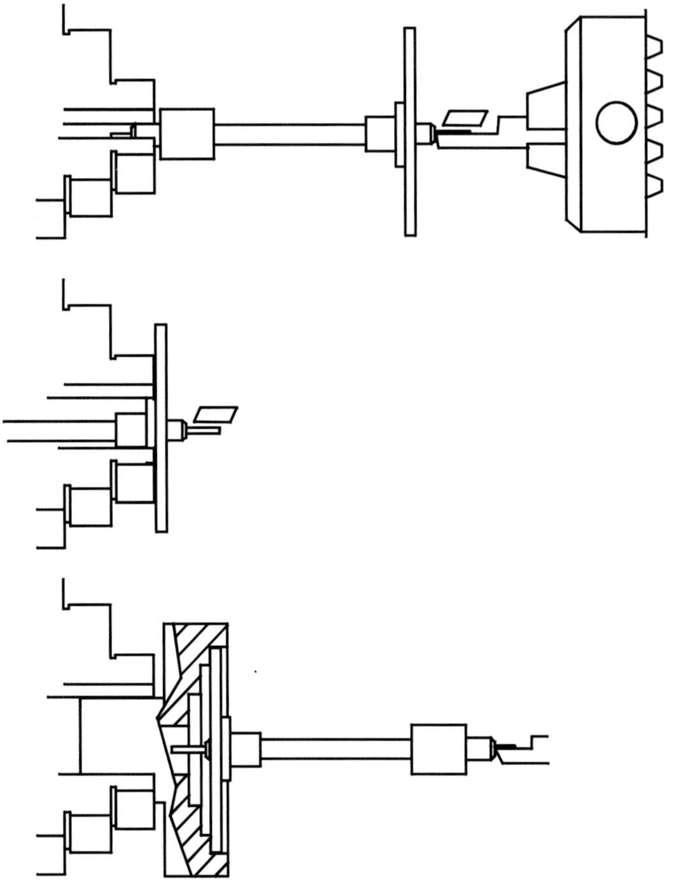

Fig. 8.32   The three ways a pivot may be presented for filing.

Fig. 8.33   *This pinion and wheel assembly has been fitted with a running bush so that it may be supported in a steady for polishing or replacing a pivot.*

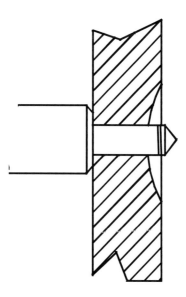

Fig. 8.34   *It is not necessary to remove lines outside the bearing surface, but allow for shake. Taper outside the pivot hole is unimportant too, but do not allow this much of the pivot to protrude as it will collect dust.*

are visible (other than those made by the act of filing). Finish filing when the pivot is a true cylinder for a length equal to the thickness of the movement plate plus the shake, or free movement between the plates. A small taper or lines outside the pivot hole are acceptable, but should not extend past the mouth of the oil cup.

The finish that is left after filing is not smooth enough for a good bearing, or running surface: it must be polished if it is a dead hard antique, or honed/burnished for a tough but not dead hard pivot. The traditional tool is a burnisher, which, as previously described, is a flat piece of very hard steel or tungsten carbide with a trapezium-shaped section, which has had 'cut' put on it by rubbing the burnisher crossways on a piece of emery paper/cloth of approximately 250 grit. This creates a surface of very fine grooves running from side to side of the burnisher and actually gives it a slight cutting action (Fig. 8.33).

Whether the pivot should be supported or not during the filing and burnishing operations, is a matter of personal judgement. If the pivot is over 1.25mm in diameter and the arbor can be held securely in the chuck of the lathe (or pistol drill or bench drill) without the work hanging out more than about 12mm then it is not normally necessary to support it. Note that it is difficult to provide support for an arbor in a bench drill (Fig. 8.34).

Do not use emery or crocus paper when supporting the pivot because abrasive dust will become trapped under the pivot, 'ball up' or aggregate and cut grooves deeper than the grit size.

Do not press very hard when filing. It is not a matter of leaning into the job with the full weight of the forearms, but of applying finger pressure. When the pivot is running free of any mechanical support, I place my forefinger beneath and use the thumb to squeeze the file against it (as shown earlier in Fig. 8.13). Since the thumb can follow the stroke of the file, this gives me an even and controlled cut. Very small pivots (below 0.75mm) cannot be safely filed in this way because the flesh of the thumb does not supply a rigid support on tiny diameters. There will be a danger of bending and ultimately snapping it.

The same constraints apply to burnishing.

Burnishing

Generally speaking, good-quality clocks produced since the middle of the nineteenth century have hard pivots that snap rather than bend. Prior to this, pivots tended to be softer and less prone to snapping off. Largely this is a matter of whether the maker of the clock was in a position to use a good-quality, homogeneous steel (crucible steel), which was more expensive than the common forged metal from a blacksmith.

The surface produced by a high-quality burnisher is both smoother and of a different grain structure to that left from filing. When viewed under a microscope (after etching slightly), annealed steel shows a surface rather like a map of small English fields – odd shapes and varying colours; these are the grains. For our purposes, the make-up of the grains is unimportant, but the different colours indicate different constituents of the steel and the grains pass small electric currents in the presence of a suitable electrolyte such as water and a small amount of salts. The currents are the result of an interchange of charged particles of carbon, iron and other components of the metal, oxides and salts present and the situation accelerates as the electrochemical differences grow greater. We call it corrosion.

If the surface is made very smooth so that even at a microscopic level there are no grooves to trap electrolyte and no discernible differences in height of the surface, the tendency to corrode would be lessened. If the grains were smeared mechanically so that the differences in chemical make-up were blurred, or even totally disguised, the tendency to corrode would be even smaller.

Burnishing smears the grains in just this way, providing a good bearing and a highly rust-resistant surface.

It is often claimed that burnishing work-hardens the surface. I find it difficult to believe, however, that any appreciable degree of hardening can be given to high-carbon steel with a hand-held burnisher used in clockmaker's fashion. The limiting stress for hardened steel, which is the pressure at which the steel deforms, is about 100,000lb per square inch (psi). Work-hardness produced by leaning on a piece of smooth hard steel must be almost impossible. However, many modern and mass-produced clocks (particularly American clocks) employ a low- or medium-carbon steel that does not have such a high limiting stress. These steels probably do work-harden to a limited extent. In any case, the 'smearing' of the surface structure of the pivots as a result of burnishing is a useful result.

A similar surface obtained with the use of an abrasive such as crocus powder or diamantine may not give the same smearing protection against rust, but clocks from the nineteenth century and earlier and quality watches were mostly polished; clock and watchmakers such as W.J. Gazeley give detailed instructions for the process. If a shaped hard steel wheel is run along the tooth spaces of a pinion with a dressing of diamantine, smearing will occur, but this is a matter of combining two forms of finishing. Not all pivots are burnished, but those in good-quality clocks often are. Modern diamond and abrasive ruby laps produce a good finish on pivots (see below).

It should be noted that the very action of the clock movement will burnish the pivots in the brass pivot holes. Consequently, if hardened pivots are seen to be free of corrosion and unscored when removed from the plates, it will not be possible to improve on the surface that has been worked on for possibly a quarter of a century. Simply touch the corner of the shoulder and the pivot diameter with the edge of a burnisher to make sure that there is no encrustation of old oil or dirt, and then take a light whisk over the surface with the very finest flour grade emery or crocus paper to make sure that every vestige of old oil has been removed there also.

If a new arbor and pivot is being made, the latter will have been turned first and then filed lightly and polished with emery or crocus paper. It is then hardened and tempered, cleaned with emery cloth (in the lathe) and finally burnished. Always use hardenable steel for arbors. There are many modern clocks that make use of low-carbon steel. They nearly all show a tendency to bend over the years, and as soon as bending begins, the tendency accelerates. I believe that many torn barrel teeth are the result of intermediate arbors that are soft (sometimes as

a result of not re-hardening after repair work), and gradually bending away from the heavily loaded spring barrel.

Mirror Finish

The finish that should result may be described as a mirror finish as viewed at a magnification of about [x]10, with no grooves or deep scratches (Fig. 8.14 again). On a hard pivot this can be achieved with abrasive polishing using a final polishing medium of 2,500 grit or finer. If the abrasive is a coated paper or fabric, it should be backed up with a brass strip, if a diamantine or other powdered abrasive is used (where the size of the grit is measured in microns), it should be applied with a bronze polisher (a flat piece of bell metal, aluminium bronze, beryllium copper or phosphor bronze). After crushing the powder with oil, it turns black and is ready to use. However, on soft steel, the best and quickest way of achieving a mirror finish is by fine polishing followed by burnishing.

Using the Burnishing Tool

A burnisher should displace the metal that it is working upon. By that I mean that the tool should traverse the metal and move the steel so that 'peaks' are rolled into the 'troughs' that can be seen at a microscopic level (magnified [x]20). Burnishing cannot be carried out on steel with a hand tool that is simply rested on the cylindrical pivot and pressed down, as there would be insufficient pressure developed by line contact between the flat surface of the tool and the cylindrical surface of the pivot to displace the metal.

For displacement to take place, the limiting stress of the steel must be exceeded. This is the stress at which the metal is no longer elastic and does not recover its form after the stress is removed. In soft steels this stress is about 40,000 psi, in hardened and tempered high-carbon steels (tempered well past purple) it is 190,000 psi. Steel that is displaced by a burnisher is work-hardened, although to what degree this takes place on a clock pivot is difficult to establish.

By traversing the tool along the pivot, the pressure is applied at the edge of the tool. This may be regarded as being point contact, where the pressures (stresses) set up are very many times greater than those that result from line contact. As far as soft steel is concerned, the practical indications are that displacement does take place. The surface, when viewed at a magnification of [x]20, shows the removal of peaks and the 'filling' of troughs. Examination of the burnisher surface does not reveal any significant removal of metal from the pivot and the pivot diameter will have been reduced by less than 0.013mm as a result of displacement.

Lubrication is almost imperative; I find that an automotive oil (10W-30) works well. I chose it because these oils are intended to withstand high pressure without the film between the metals breaking down. However, the metal that the pivot is made of will have an influence, and the lubricant that has worked on a dozen different pivots may perform badly on the next. Be prepared to switch to another oil (or even spit) if the burnisher refuses to produce a black polish and leaves the pivot grey. Burnishing is an art, and even with a perfectly prepared burnisher, you will almost certainly have to practise on a number of pivots before you get the hang of it. It is unlikely that a hand-held tool can truly burnish hardened steel (although industrial burnishers with supporting rollers certainly can), but this is a personal opinion.

Making Your Own Burnisher

The burnisher that was used for the pivot that is shown in Fig. 8.13 was made from tungsten carbide and the materials used are shown in Fig. 8.35:

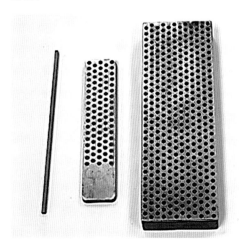

Tungsten carbide blank $^3/_{32}$in $\times$ $^1/_8$in $\times$ 5in, #04120630, purchased from MSC (left)
Diamond sharpening 'stone' extra fine, #05126594 (middle)
Diamond sharpening 'stone' coarse, #05126230 (right)

The coarse whetstone on the right of the photograph in Fig. 8.35 is one that we happened to have in the workshop and is much larger and more expensive than the one listed above. The tungsten carbide blank is unground and needs to be surfaced all round on the coarse diamond whetstone; the edges must be treated in the same fashion although the radius produced should not be greater than about 0.125mm or the pivot will have a significant radius at the shoulder when

ABOVE: *Fig.8.36　Applying the burnishing tool.*

LEFT: *Fig. 8.35　The materials needed to make a tungsten carbide burnisher.*

finished. Since a burnisher cannot conveniently be used right up to its ends, it is sufficient to dress the surfaces until none of the original surface shows, for about 75–100mm in the middle. The diamond whetstones are made of aluminium sheet that is pierced by rows of holes filled with the diamond grit and a plastic binder.

Place the blank on the surface of the stone so that it lies at an angle of 45 degrees, press down lightly (it is likely to roll over if you press too hard) and stroke it obliquely across the diamond surface so that the rows of diamond do not pass in sequence over the same place on the blank, producing a rippled surface. Try to make sure that the whole of the tool surface contacts the whole of the whetstone surface at each stroke – the tool surface is the middle portion (75–100mm) that you are dressing for use. In Fig. 8.37 you can see that one side has been dressed on the coarse stone to within about 18mm of the ends; note also that I have ground a mark on this surface so that I know which one it is.

Make sure that the blank is completely flat

over this area and free of blemishes. In Fig. 8.37 there is a blemish remaining on the surface, right on the edge of the blank; this must be removed before use.

All four sides must be dressed on the coarse stone. When this has been done, run your thumbnail down the edges and examine them with a magnifying glass. If there are any small notches, they must be removed before finishing the tool, because these will scar the pivot and produce grooves.

Tungsten carbide blanks are made from powder (a mixture of the carbide and a 'binder' that is usually nickel, or a nickel alloy), which is heated and compressed. As a consequence, the blank may well have places along the edges where particles of the carbide are 'missing'. Unless the line of the edges is continuous, the missing particles turn that edge into a saw. If the removal of a notch is clearly producing a radius that is much greater than the 0.125mm stipulated, remove it by coarse stoning the surfaces that are not going to be used for the polishing and

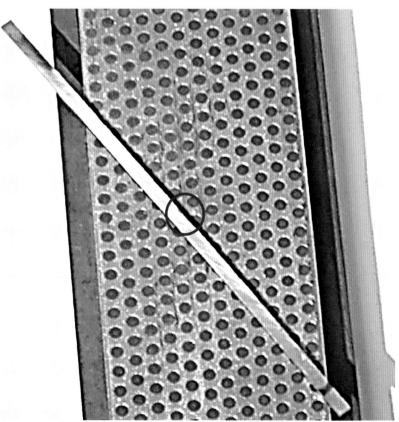

Fig. 8.37   Try to make sure that the whole of the tool surface contacts the whole of the whetstone surface at each stroke. The tool surface is the middle portion that you are dressing for use. Here you can see that one side has been dressed to within about 18mm of the ends. Note that a mark has been ground on this surface so that it can be easily distinguished and there is a blemish (circled), which must be eliminated.

burnishing, and then touch up the edges again with the fine diamond whetstone.

The side opposite the one bearing a mark is to be the finishing, or burnishing, side. Grip the smaller fine grit whetstone in a vice (it will not stay still on the bench as the big one does) and stroke the tool over it in the same manner as before (see Fig. 8.38). Make sure that all the coarse strokes are removed on this side and make a final check of all edges with a thumbnail and a magnifying glass of at least [x]5. Correct them with the fine stone if necessary. Let me emphasize again that faulty edges are usually the cause of any grooves that appear in the finished pivot, and you can very easily ruin a good surface with a bad edge. When using your burnisher, take care – tungsten carbide is brittle, though not disastrously so, and dropping the tool onto another hard surface may chip the edge; wrap it in cloth when not in use.

Keep the whetstones clean by washing in soapy water, dry completely and do not apply any oil or other cutting or cleaning fluid.

HARDENING AND TEMPERING

Hardening and tempering of pivots is a simple matter of heating to red heat and plunging the steel into oil. If silver steel or some other non-distorting steel is used, you simply need to ensure that the pivot enters the cooling oil end on, not broadside. If it went in sideways on, there would be a strong chance that the metal would bend as one side entered the cold oil first and shrank. Other high-carbon (or hardening) steels must be treated more carefully. Take some soft iron wire, such as is used in flower arranging, and carefully wrap it all around the arbor, pinion and, even more carefully, the

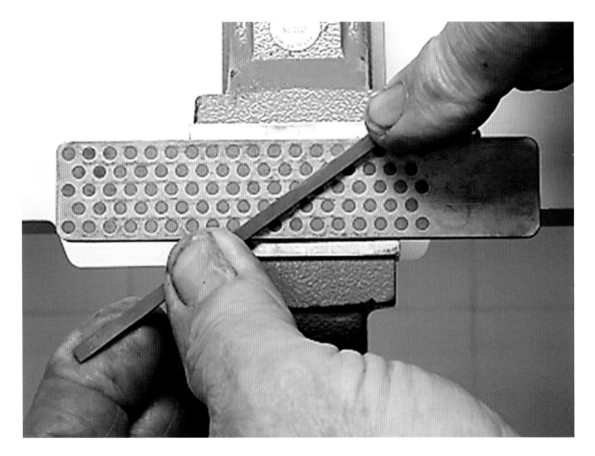

Fig. 8.38   For the burnishing surface, grip the smaller fine grit whetstone in a vice and stroke the tool over it.

pivot. The purpose is to add mass to the piece so that a more even cooling will take place at the centre of the mass where the arbor lies. It is still necessary put the pivot in the oil end on. The degree of hardness obtained is likely to be a little less than when the arbor is unclad, but since the metal is to be tempered, this hardly makes any practical difference.

Tempering is a matter of raising the hard steel to a known temperature and no further. The rate of heating and cooling is of no importance to the clockmaker. Four methods of obtaining a known temperature are commonly used:

- A controlled muffle with pyrometer to measure the temperature
- A mixture of salts or low-melting-point metals that are compounded to melt at the required temperature
- Cleaning the metal until bright and then carefully heating until one of the temper colours is obtained (in this case, blue). For blue temper, the piece can be dipped, uncleaned, into old motor sump oil and warmed until the flame of the burning oil is just self-sustaining and will continue for a few seconds after the heat source has been withdrawn. This last method works well for relatively large pivots such as longcase and bracket clocks, but not for carriage clocks or the normal round French movement found in mantel clocks.
- Heating in a domestic electric cooking oven, particularly one with a self-cleaning function. The tempering colours used for pivots and pinions are covered by the range from dark brown to dark purple. The former represents a temperature of 260°C (500°F), the latter 280°C (536°F).

An alloy for the tempering bath for the second method can be made by adding small amounts of soft solder to lead. Since the solder is itself an alloy of lead and tin, which is probably not defined by the manufacturer, this is very much a matter of trial and error. Less than a tenth of the alloy will be made up from the solder.

Prepare a test piece of steel, brightening the surface with emery paper and making sure that there is no grease or oil on it. Melt common lead in a container that is fairly robust (a tin can will not do) in an area that is well ventilated. When the lead has melted, adjust the temperature so that no bubbles or violent disturbance is seen, and add small amounts of solder to the melt. You will need to adjust the heating again because the solder lowers the melting point. Test the temperature of the melt by dropping the steel test piece onto the surface – it will float and show colour changes clearly. If there is not enough solder in the mix, the tempering colour will pass purple and reach silvery blue. In this case, take the steel out, brighten the surface and add a bit more solder before floating the test piece again. Once the alloy has been established, make a note of the proportions of lead and that particular solder, so that the alloy can be made up more quickly next time.

I must emphasize the need for good ventilation, as lead vapour is very harmful, although you will keep it to a minimum if the melt is not boiled. Salt baths are also harmful and tend to produce nitrogenous vapours very readily. (Face masks are helpful but will only stop vapour or fine droplets, **not** gas.)

Although I have detailed the making of a lead/tin tempering alloy, it is really only needed for the more difficult tempering jobs such as bluing hands. For pivots, the simple technique of brightening the steel and then slowly heating the arbor until the colour appears, is both quick and all that is necessary. If the colour is exceeded, you must go back to fully hard again by heating to red and quenching in oil.

Chapter 9

Removing and Mounting Gear Wheels

In clockmaking parlance, gear wheels are simply referred to as 'wheels' and any gear with more than twelve teeth is usually counted as being a wheel. Gears with twelve teeth or fewer are referred to as pinions. There are several reasons why a wheel may be taken off the spindle or arbor of a clock movement:

- It is removed deliberately to carry out a repair to the pinion or the pivot.
- The wheel is damaged and needs to be repaired or replaced.
- Ill-treatment has resulted in it becoming loose.
- The wheel has been removed and replaced eccentrically.

- The pivot is worn or damaged and repairs require the wheel to be removed first.

Before re-pivoting an arbor, it is often necessary to remove the wheel. Thoughtful clockmakers solder the collet that the wheel is mounted upon to the arbor. Such a collet only needs a little heat and the wheel and collet drop off quite nicely. However, some makers and repairers had a habit of driving the collet onto the arbor, and these cannot be easily removed without risking damage to the arbor. They should be turned off in the manner shown in Fig. 9.01, first releasing the wheel by removing the part of the collet that was swollen by punching, and then totally turning away the rest of the collet.

A   B

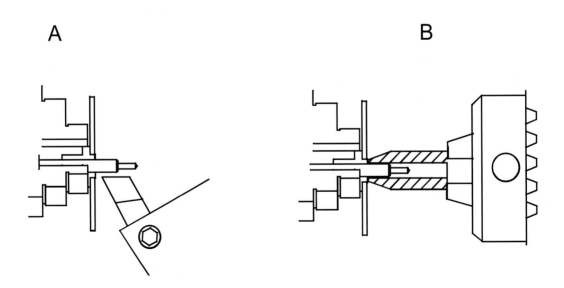

Fig. 9.01   First cut away the swaged (riveted) portion (A), then make a steel bush similar to a ring punch (B). Make sure that it does not touch the wheel bore before using the tailstock hand wheel to press it out.

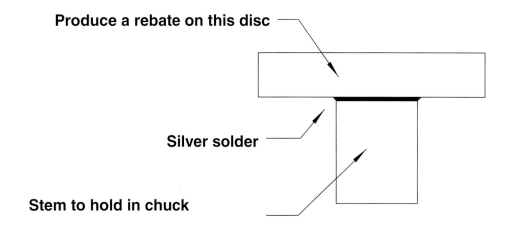

Produce a rebate on this disc

Silver solder

Stem to hold in chuck

Fig. 9.02　Make a simple mushroom from a rod of brass (about 12.5mm in diameter).

If the wheel is being taken off because it is scrap, there is no problem. Take all the relevant dimensions, snip through the crossing-out to remove the rim and then turn down the hub on the wheel and the collet. If there is sufficient diameter of metal in the old collet to allow turning down to the diameter of the hole in the new wheel, then it will not be necessary to turn it all away. Fig.09.01B shows the wheel being forced off the old collet by the tailstock chuck.

I have been told that driven-on collets will fall off if heated to dull red two or three times. This has never worked for me, however, and I do not like the idea of over-heat-treating elderly steel in this fashion. The wheel must be removed first, of course.

TRUING WHEEL BORES

The wax chuck was described earlier in Chapter 8. This method of gripping the work can be varied or simplified if the wheel has been taken off the collet and is to be set up for boring. Make a simple mushroom from a rod of brass (about 12.5mm diameter) and a disk that is a little larger than the wheel. Set the mushroom in the three-jaw chuck and use a small gas torch to warm up the disk. You want it to be hot enough to melt a stick of shellac or hard wax.

The traditional method would then be to spin the chuck and apply the shellac to the disk. Stop the chuck and set the wheel as near concentric with the centre as you can, keeping the shellac in the plastic state by warming it from time to time. Now spin the lathe and use a cocktail stick to lightly touch the outside diameter, gradually moving the wheel over until the outside runs dead true. Be careful: if the point of the stick catches in the gear teeth, it will flick the wheel off and you will have to start again. When the wheel is running true, leave the lathe spinning until the shellac is hard, and then bore out the centre to the diameter required using very light cuts.

I find this technique difficult and so I produce a mushroom with a recess bored to closely fit the outside of the wheel and I use Loctite or a similar adhesive to hold the wheel in place. I have shown one being used to file a pivot in Fig. 8.32 in the previous chapter; this is a tool with several uses.

REMOUNTING WHEELS

Wheels are carried either on a special seating called a collet, or directly on one end of the pinion. To obtain true running of the wheel, the bore must be concentric to the outside diameter and the seating that the wheel will be

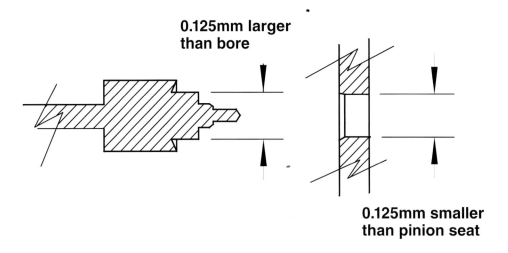

0.125mm larger than bore

0.125mm smaller than pinion seat

Fig. 9.03   If the wheel is to go directly onto the pinion, the pinion is machined to accept it. The hole in the wheel is commonly smaller than the newly turned part of the pinion so that a drive fit is obtained.

pressed onto must have been machined true to the pivots. If the wheel is to go directly onto the pinion, the latter is machined to accept it (Fig. 9.03).

The hole in the wheel is commonly about 0.125mm smaller than the newly turned part of the pinion so that a drive fit is obtained. If too much is left on the pinion, there is a danger of the wheel wandering to one side or the other

as the pinion cuts its way in. A small chamfer at the beginning of the seat on the pinion will tend to prevent too fierce a cutting action but it can still wander slightly, so check as the operation proceeds.

Attention must be paid to the details of the corner of the seat and the edge of the hole in the wheel so that the wheel will sit right back against the shoulder prepared for it. With the

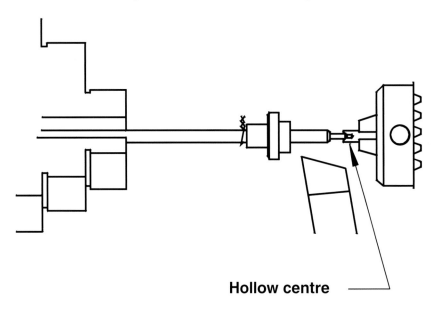

Hollow centre

Fig. 9.04 Mounting the collet by soldering a roughly turned brass collet at a predetermined position on the arbor, and then turning it to suit the bore of the wheel.

wheel in position, it only remains to lock it there. One method is to use a ring punch that is only 0.25–0.5mm smaller than the wheel hole, support the back of the pinion and tap the slightly protruding leaves of the pinion with the punch, swaging them over. However, if the drive fit is good, it should be sufficient to use a product such as Loctite to put the finishing touch to the job. Use just enough to be retained inside the wheel bore, as you do not want to have to clean the pivot leaves afterwards.

Mounting the collet is usually a simple matter of soldering a roughly turned brass collet at a predetermined position on the arbor. (A twist of fine wire will mark the position and stop the collet sliding away.) The arbor is then set up either using a hollow centre to support it directly in the chuck, or a lathe collet and ensuring that the pivot runs true.

When this has been achieved, the collet is turned to produce a diameter and shoulder to seat the wheel neatly. A nice push fit is required, so that the wheel just pushes on without strain and without any sideways shake. Once in position, clench the wheel firmly by swaging the end of the collet as shown in Fig. 9.05, or use a ring punch again. The former is best, as it is very easy to ruin the setting of the wheel upon the collet by hitting the ring punch out of square.

When all is done, the wheel must run without wobble in any plane when it is spun between hollow runners supporting the pivots; the depthing tool is useful for checking this.

Alternative Mountings

There are other methods of mounting wheels on arbors. The most common is the use of a washer with a 'keyhole' in the centre and an arbor with a machined groove. Keyhole washers (also called friction disks) are used to allow the barrel arbor to rotate within the great wheel (or going barrel) for winding. The wheel is free to rotate on the arbor and bears against the end of the largest diameter of the arbor. The spring washer is pressed down against wheel and is locked in place by sliding into the groove (Fig. 9.06). This produces friction between the arbor and the wheel; a small pin holds the washer and wheel together, preventing the washer from sliding out of the groove again.

This is a very useful method of fixing the great wheel in position, but the pin that locks wheel and washer can be troublesome. Often it is left standing about 1.5mm proud with the intention that it should be pulled out by gripping with a pair of pliers. This is not always successful and, after two or three abortive attempts to remove the pin, there will be nothing left to grab hold of. However, if there was a hole in the end of the barrel that could be lined up under the pin, the latter could be driven through quite

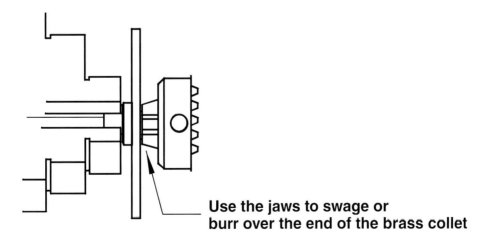

**Use the jaws to swage or
burr over the end of the brass collet**

Fig. 9.05   Once the wheel is in position, clench it firmly by swaging the end of the collet.

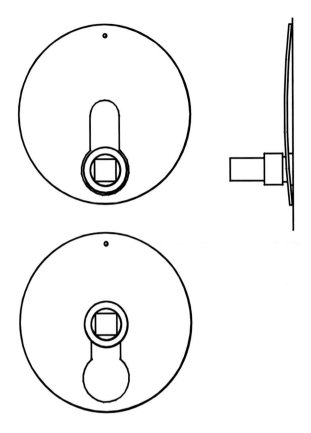

Fig. 9.06   The wheel is free to rotate on the arbor and bears against the end of the largest diameter of the arbor. The spring washer (keyhole or friction disk) is pressed down against wheel and is locked in place by sliding into the groove.

readily (Fig. 9.07). There is often a hole in that end of the barrel so that the line can be knotted and the knot pulled inside the barrel, but more often than not it does not correspond with the position of the pin.

A much better method of locking the keyhole washer in place is to drill and tap the wheel so that a small screw can be inserted. Since there is no reason for the head of the screw to bear down upon the surface of the washer, a hole that allows the screw head to pass through and lock against the wheel is permissible. It has the advantage of reducing the number of parts that could conceivably rub against the inside of the clock plates.

Large clock wheels are sometimes fastened to the collet with screws, and most tower clock wheels are fastened in this fashion. The only problem that this poses is in re-tapping a hole that has stripped its thread: the repairer will need a long-reach tap wrench. A variation on this is to drill clear holes through both wheel and collet and fit bolts (nut and screw) to hold them together (Fig. 9.08). I can see no necessity for using screws or bolts to

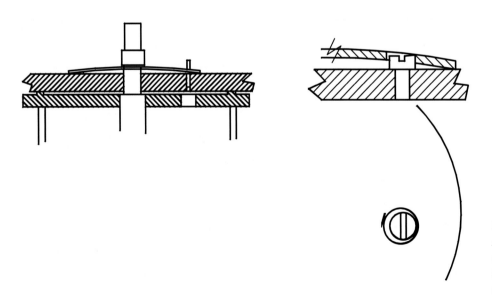

*Fig. 9.07
A locking screw does not need to bear down on the spring.*

Fig. 9.08 A variation is to drill clear holes through both wheel and collet and fit bolts (nut and screw) to hold them together.

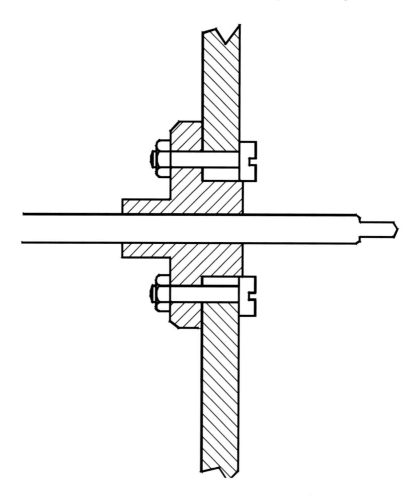

hold smaller wheels on their collets if the latter is swaged to grip the wheel as described above. Traditional clockmakers' lathes are not beefy enough to employ this system and I believe that the use of screws on high-quality clock wheels was a method of avoiding the need to use a ring punch for locking the wheel on its collet.

If a clock wheel is originally mounted on the arbor, its removal will probably rule out any hope of remounting concentrically without boring out the wheel slightly. (A remount of this kind can be achieved sometimes, but you cannot count on it.) Once it has been bored out, the seating is of course too small for the wheel. The reason for mounting the wheel in this manner – keeping the high gear loadings in one movement plate – and the restricted space that is common, often rule out the possibility of turning a new seating further along the pinion.

If this is the case, prepare a thin bush of brass, with a bore that can be pressed over a newly turned diameter on the pinion and an outside that is larger than the wheel bore (Fig. 9.09). Once this is done, face back the pinion with very light cuts, or even with a medium-cut file, so that the leaves do not get bent, and produce a seat that the bush can be forced onto. The face of the drill chuck is utilized for the force fit operation. When the bush is in place, check that the pivot is still running true and then turn the outside to fit the bore of the wheel. Leave a tiny shoulder, no more than 0.125mm wide, for the wheel to fetch up against.

There is not sufficient metal in this bush to employ the swaging technique, so simply clean the surface of the bore and use an industrial adhesive. Fit the wheel and spin the lathe chuck under power. A light touch with a forefinger

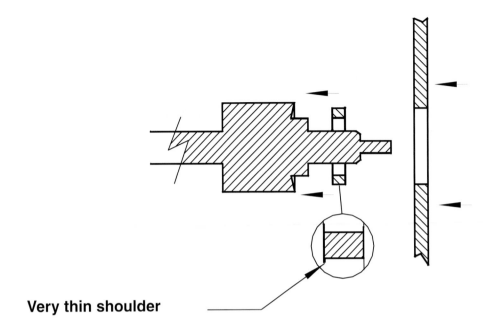

Very thin shoulder

Fig. 9.09   *Prepare a thin bush with a bore that can be pressed over a newly turned diameter on the pinion and an outside that is larger than the wheel bore.*

will set the wheel true so that there is no wobble, and the lathe is left spinning until the adhesive has set – between 5 and 10 minutes.

MAKING A PINION AND PIVOT FROM NEW

It is not always possible to re-pivot an arbor, and so it must be made from scratch. This is the case also when a pinion develops a broken leaf.

Gear cutting is not a subject to detail here and I find it more economical in any event to buy pinion blanks from specialist makers and either fit those (as heads) or turn them (complete arbors).

The pinion blank is usually a short length of high-carbon steel capable of being hardened and tempered. If there are eight or more leaves there will probably be enough diameter of solid metal beneath the tooth root to allow the drilling of a hole clear through. In this case, the pinion can be made as a head only (Fig. 9.10). The arbor is machined at one end to produce a

pivot and at the other to make a cylinder that fits the drilled hole in the pinion head with the pivot protruding beyond it. Pinions with seven or fewer leaves hardly ever have a sufficiently large root diameter to allow drilling for the arbor.

Eight or More Leaves

If there is no reason to preserve the original arbor, it is always best to make the arbor and pinion as one piece. This involves turning the plain portion (the arbor) of the replacement pinion stock and reducing the length of the gear form (the leaves) to obtain the length of pinon needed. Before turning the new pinion stock, use a fine-toothed saw to make a cut down the leaves to the root. This marks the extent of the turning for the arbor. Facing the leaves always runs the risk of bending them; making a cut like this reduces that risk and means you only needs a very light cut to clean the sawn surfaces at the finish.

As Fig. 9.11 shows, the machining of the

Fig. 9.10 A separate head is possible with eight or more leaves (top). Smaller gears should be make integral with the arbor unless the teeth are very shallow (bottom).

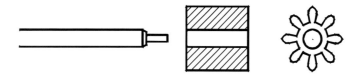

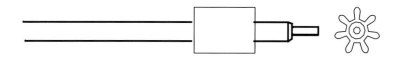

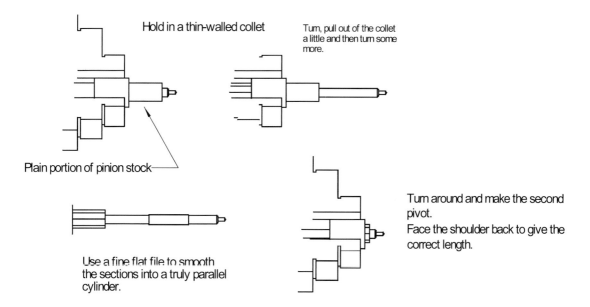

Hold in a thin-walled collet

Turn, pull out of the collet a little and then turn some more.

Plain portion of pinion stock

Use a fine flat file to smooth the sections into a truly parallel cylinder.

Turn around and make the second pivot.
Face the shoulder back to give the correct length.

Fig. 9.11 The machining of an arbor must be done section by section.

arbor has to be done section by section; the pivot is tackled first and a small length of the arbor diameter and then the work is pulled a short distance off of the chuck and the next section is machined to turn the next section and so on. The stock is held using a thin-walled bush and the various sections are finished with a fine flat file to produce a clean cylindrical arbor of constant diameter. When the arbor is finished, the work is turned around in the bush and the second pivot is turned (Fig. 9.11).

As already mentioned, a pinion with at least eight leaves can be repaired by simply making a pinion head, machining the old pinion off and producing a cylinder for the head to slide onto (if there is a reason to preserve the old arbor) or making a completely new arbor from rod of the right diameter. I prefer the latter solution for two reasons: it avoids the necessity of turning a long, slender arbor by sections, and if the pinion stock is long enough, more than one head can be cut from it. It is not uncommon to find that a pinion for one clock suits another at a later date. Pinion stock is expensive.

Machine the head by sawing a short length off the pre-formed pinion stock with a fine-toothed saw – 32 tpi appears to be the finest available. Hold it in a brass collet and face both ends using light cuts and a very sharp tool. Start the hole with a centre drill; I always find that I get the best results from using a drill straight out of the packet and drilling right through. Run at as high a speed as the facing tool will stand and with very light cuts because the leaves will then be less likely to bend – unless the tool loses its point.

Sometimes you may find it more convenient to turn a pivot on one end of the pinion head and then bore the other so that it slips over the old arbor (Fig. 9.12). Since it is not important for the arbor immediately behind the head to be concentric with it (as long as it carries no working parts), you will be able to simply file

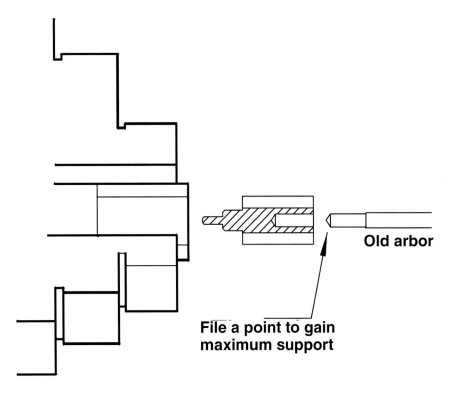

Old arbor

File a point to gain maximum support

Fig. 9.12   A pinion with at least eight leaves can be repaired by simply making a pinion head and machining the old pinion off and producing a cylinder for the head to slide onto, or making a completely new arbor.

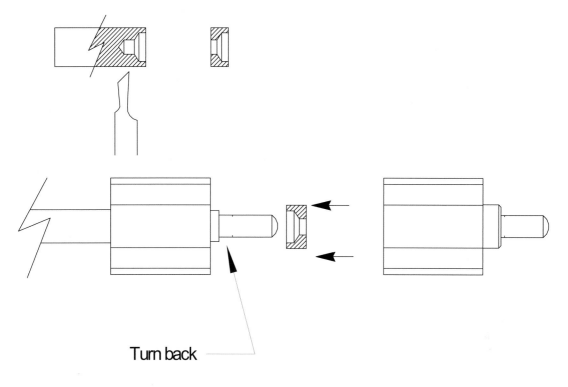

Turn back

A false shoulder may be fitted when there is not enough arbor to form a shoulder.

Fig. 9.13   Adjusting the shoulder between pivot seat and pivot.

it in the lathe until the arbor slips into the hole prepared for it in the head. The hole should be twice as long as its diameter; anything less, and the head is at risk of falling off in service. A bit more stability will be gained if the end of the arbor is filed to match the end of the hole.

After machining, the pinion head can be hardened and tempered in oil, and the arbor and pivot also. The job will be easier to manage because tempering (and hardening) is adversely affected by large changes in section such as would result from a one-piece pinion, pivot and arbor. Separate pieces make for more readily controlled hardening.

Degrease both parts thoroughly and use a cyano-acrylic adhesive to fix the head in position. Polish pivots and pinion leaves before assembly.

If the shoulder between pinion seat and the

pivot is small and does not appear to be suitable, make a false one in steel and slip it on; lock with an adhesive and, when set, use a pivot file to remove excess in the corner of shoulder and pivot (Fig 9.13).

Seven or Fewer Leaves

Pinions with seven or fewer leaves may need to be made in one piece. This will require normal turning practice but life will be simpler if you set about the various tasks in a logical order that reduces the number of times the work has to be moved in the chuck. Fig. 9.14 sets out a good sequence. Machining starts on the arbor with the pinion head end in the chuck. The reduction of the main diameter of the arbor is carried out in stages and a file is used to smooth these out later. Since the collet for the wheel is always turned true to the pivot, any slight inaccuracies

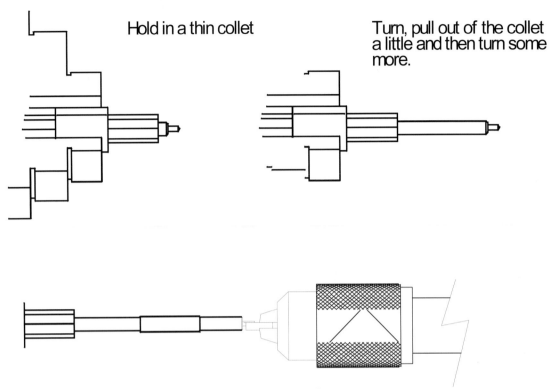

Hold in a thin collet

Turn, pull out of the collet a little and then turn some more.

Use a flat file and support to smooth out the different diameters.

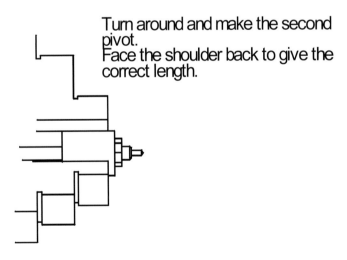

Turn around and make the second pivot.
Face the shoulder back to give the correct length.

Fig. 9.14   The sequence for making pinions with seven leaves or fewer.

in the arbor itself will not matter. Polish before and after hardening and tempering.

BORING OUT SMALL HOLES

Consider a pinion that has been drilled out to fit on a new arbor and (as happens very frequently), the drill has wandered away from the centre line; there is no way that you can persuade the wheel and pinion to run true after this happens. So the small hole in the pinion needs to be bored out in the lathe.

There is a limit to the 'smallness' of hole that can be bored with normal workshop tools (excluding spark erosion, laser cutting and so on) – the fly pinion on a carriage clock, for instance – and the only solution for a normal clockmaker's workshop may well be to make or buy a complete pinion and arbor.

I have two tools for drilling small holes. The first (Fig. 9.16) is really a one-sided drill and it must be made no longer than is necessary to keep it stiff; it is really only suitable for boring gear wheels or anything that is no thicker than 2–4mm. The second tool (Fig. 9.17) is simply a scaled-down boring bar, which is easily made on the lathe.

Fig. 9.15   The chuck is holding one end of the arbor while the steady supports the other end. It is not necessary for the chuck end to be held absolutely true when the pivot is being corrected or replaced.

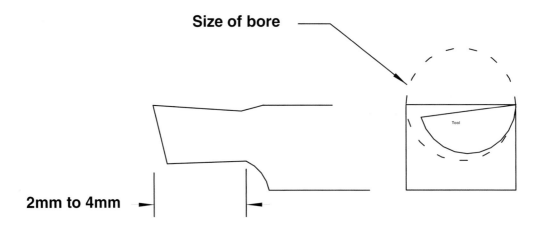

Fig. 9.16   *Relative sizes of a boring tool and the smallest hole that it will bore.*

Fig. 9.17   *This is a scaled-down boring bar that is easily made on the lathe.*

Making a Small Boring Tool

The boring tool is made from a piece of drill rod about half the diameter of the hole to be bored. Whilst it is soft (and preferably red hot), the metal is bent through 90 degrees and allowed to cool. Using a saw and a file, the end of the 'dog leg' is formed to make the standard boring bar form. It is then hardened and tempered to very light amber. The holder for the bar is a piece of mild steel that has been clamped in the tool post and drilled from the three-jaw chuck to fit the boring tool. As a result it is dead on centre.

A saw cut (using a small hacksaw or jeweller's saw) is made above the drilled hole and just cutting into it, allowing the holder to spring slightly. When the holder is put into the tool post, one of the screws will press down on it and close the holder onto the boring bar. By slackening off the first tool post screw, the boring bar can be rotated by hand to place its point slightly above centre; this means that the cut will be relieved if the tool bites too deeply. It saves the tool from breaking. (Pay no attention to the two spotted holes on the top of the holder in Fig 9.19. It was a piece of scrap metal that I picked up quickly to make my tool from; the holes did not interfere

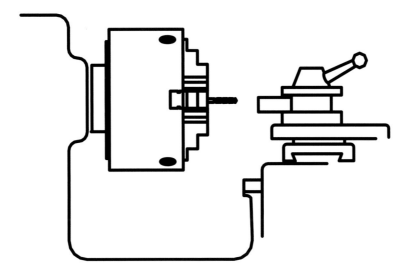

Fig. 9.18a   Drill a piece of rectangular section bar for a convenient silver steel rod (2mm).

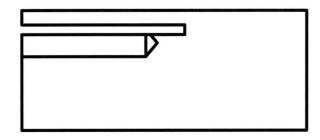

Fig. 9.18b   Cut a slot with a saw and insert the tool. It is held in place by tightening the tool post screws.

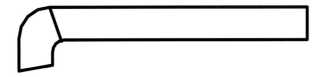

Fig. 9.18c   The tool is made by bending silver steel while red-hot and filing the bend to suit.

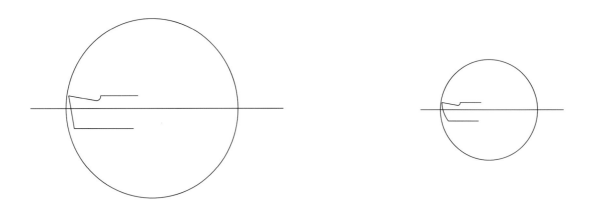

Fig. 9.19   Left: when boring, the tool point may be placed above centre and will then relieve itself if the cut is too deep and the bar dips. Right: curve the front clearance for small bores.

Fig. 9.20   Use a split bush or a collet to hold the pinion.

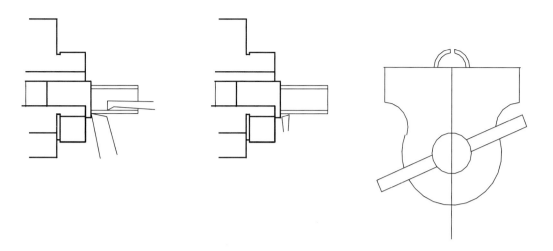

Fig. 9.21   The chuck is holding one end of the arbor while the steady supports the other end. It is not necessary for the chuck end to be held absolutely true when the pivot is being corrected or replaced.

Fig. 9.22   A wheel that has been mounted on a roughly turned rod is being set in a wax chuck.

with the tool so I did not waste time tidying it up.) This tool is incredibly useful for truing up drilled holes (fig 9.20) or the holes in wheels that had been previously pressed onto pinion leaves.

Boring the Hole

I use a split bush (Fig. 9.21) to hold the pinion, but if you have a collet that is even better. The bush is turned and bored to suit the pinion and parted off the bar at one setting, so there is no question but that outside and inside diameters are concentric. With a jeweller's saw I make a slit in the bush. After removing any burr from the saw cut, it is mounted in the three-jaw chuck with a small amount sticking out so that I can check for concentricity when running (see Fig. 6.08).

The pinion and wheel are now slipped into the bush and the chuck is tightened and then checked to see that all is still running true.

Fig. 23   The rough rod has been cut and a pivot and shoulder turned from it.

Chapter 10

Fly Cutters

Fig. 10.01 shows standard holders that are available from tool suppliers and that accept a piece of square silver steel. I prefer this form of HSS tool bit because a square bar can be held more firmly than a round one. The tools shown are intend for facing a work piece or similar work; another model, with the slot at right angles to the axis, is used for gear cutting.

Fig. 10.02 shows a DIY variety that can be held in the lathe mandrel or a Morse taper adapter and that accepts both square- and round-section silver steel bar. If you are using round bar, file a flat on the bar before starting any of the follow-

ing, to give it a more positive, repeatable location. The bottom of the holding screw should be square, or a hollow cone so that relocation of the flatted part of the tool presents no problems. Use silver steel for the cutter: it is easy to harden and its tendency to be brittle does not affect fly-cutting at high speed – 2,000rpm or higher. Note that the slot or hole for the tool is not set on the centre line of the holder or whatever means is employed to rotate it. It is offset so that it is the edge that lies on the centre line. The reason will become clear below.

Fig. 10.1   Standard holders that accept a piece of square silver steel are available from tool suppliers. rdgtools. co.uk

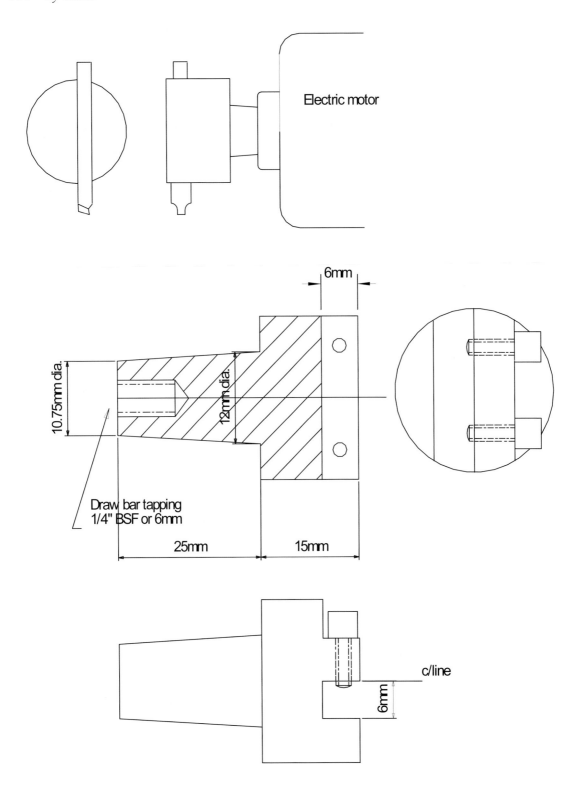

Fig. 10.2   A DIY variety of holders that can be held in the lathe mandrel or Morse taper adapter for both square round bars.

MAKING A GEAR-CUTTING TOOL FOR THE FLY CUTTER

Materials needed:
- Silver steel bar or tool steel bar of either rectangular section or round (10–15mm × 10mm or 15mm)

Tools needed:
- Parting tool with a semicircular end
- Holder for either rectangular tools or round
- Container of dry chalk or dry sand
- Gas torch capable of heating steel to bright red heat
- Pair of tongs or pliers

Stage 1: Annealing

Silver steel or any high-carbon steel is most likely to be quite tough as received; it is best to anneal it.

Hold the bar of steel in tongs or a pair of pliers that you do not mind ruining and heat to a bright red at one end (a length of about 20mm) and maintain that temperature for 2 or 3 minutes then quickly plunge it into the sand or chalk with about 100mm surrounding it on all sides. This is to ensure that the metal cools down very slowly so do not take it out of the insulating material until at least half an hour has gone by.

Do not be tempted to lift it out of the can with your fingers – it will still be very hot.

The steel will now be a great deal easier to machine.

Stage 2: Grinding

Check the form and size of gear tooth you will be cutting and note the radius of the tooth top – it will be $1.57 ×$ the module of the wheel. Select or grind a parting tool with a blade twice this width and then grind the end into a semicircle. This is a relatively easy thing to do by eye but if you prefer to check the result, drill a hole of the same radius as the semicircle and saw or file it in half.

Stage 3: Machining the Cutter

Lock the silver steel bar in the holder and make

sure that it is secure, because the machining that is about to take place is an 'interrupted' cut. Turning only takes place for a small part of the circle and the tool will hit the work piece with a thump at each revolution.

Lock the parting tool in the tool post with its top surface on the lathe centre. This is a tough job and I would recommend a top rake of 5 degrees and front relief the same to bolster (support) the cutting edge. After machining across the end of the silver steel bar to remove saw marks, position the tool to cut down one side of the bar (Fig. 10.03) to produce the form of the cutter. When the depth matches the dra ing, withdraw the parting tool and move it sideways by the desired width of the cutter and cut down the other side to produce the finished form.

The metal on either side of this form should be cut back so that, when the cutter is used, only the tooth form touches the wheel blank and cuts the space between adjacent teeth. This relieving should leave the cutter just as wide as the pitch of the gear teeth – the pitch in this case being the distance between the tips of adjacent gear teeth.

Stage 4: Hardening

Take the new fly cutter out of its holder and file a top rake; this will remove the dimple where the impact of each revolution took place and produce a sturdy cutting edge. Make the angle of rake 10 degrees relative to the centre of the holder. A steeper rake on a fly cutter clears the chips or swarf away from the cutting edge, but only if free-cutting brass is used (70/30 brass with no lead will stick).

It may now be hardened. Heat to bright red heat and allow it to 'soak' in the flame for about a minute and then plunge it into a can of oil (any old oil) and let it cool down until you can handle it. Clean it of oil and finally use an emery stone on the top rake to restore the cutting edge.

Heat the other end of the bar and whe the bright steel of the rake begins to turn yellow, quench in water. This will improve its ability to take the shock of cutting the gear.

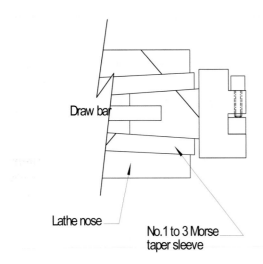

Draw bar

Lathe nose

No.1 to 3 Morse
taper sleeve

Attachment of fly cutter holder to the headstock of the lathe.

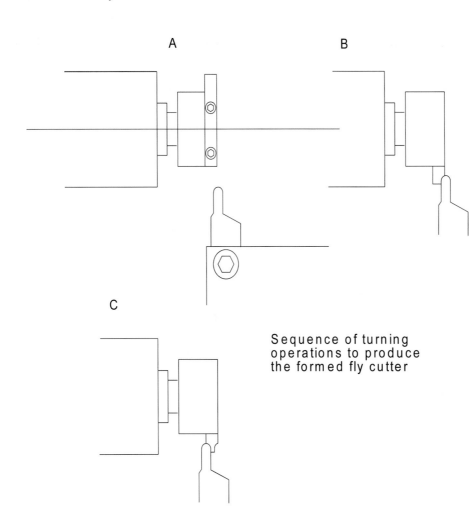

A

B

C

Sequence of turning
operations to produce
the formed fly cutter

*Fig. 10.3
After machining
across the end of
the silver steel
bar to remove
saw marks (A),
position the tool
to cut down
one side of the
bar to produce
the form of the
cutter (B); then
repeat on the
other side (C).*

Stage 5: Finishing

This is simply getting the tool ready to machine gear teeth. Remembering how the tool was held in the holder, turn it end on end and lock it in place with the rake on the centre line. The arcs that were cut in making the gear form had a centre that was also the centre of the holder. Now that has changed, and the cutter has relief in a tooth form that is constant from top to bottom. In other words, as the tool is sharpened over the years by grinding down the rake, the form will remain unchanged and will still cut an accurate gear form. Until I gave up practical work and sold my workshop, I had the first of these cutters that I made in 1974 and it was still usable in 2019. It is a very robust tool.

In Fig.10.04 all the curved surfaces are relieved.

Fly cutters only work well when cutting through relatively thin materials, such as train wheels and greatwheels (we are talking about the use of fly cutters for gear tooth forming

only). I have never had any success in cutting pinions in this way.

Use the wheel cutting tables for standard gears to establish diameters and depths of teeth. These are available from Thorntons (Successors) Ltd and other tool suppliers.

Abide by the following safety instructions when using a machine:

- Wear goggles or a full face mask to protect the eyes from particles that bounce from the work or any part of the machine. A full face mask will protect the skin of your face as well.
- Do not attempt to clear swarf away with your hands; use a hook and, if the handle is a loop, do not put a finger through the loop. One of my fellow apprentices in 1948 did that when milling a large surface; the brush was caught by the tool and his hand followed it! One month later after surgery he did it again, after that he took a job as a window cleaner.

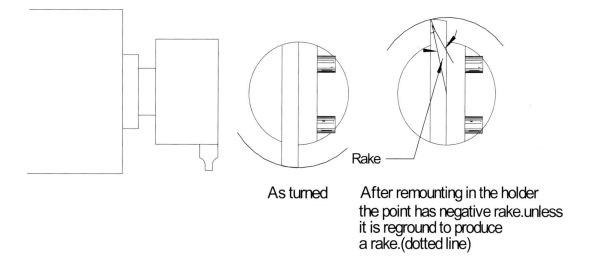

As turned

Rake

After remounting in the holder the point has negative rake.unless it is reground to produce a rake.(dotted line)

Fig. 10.4   The rake is produced by turning the tool over.

Chapter 11

Flat Depthing Tool

This is a fairly simple tool to make. It does not require much use of the lathe but it is such a useful instrument that I have included it here. The body can be made from steel or brass plate because the wearing parts are hardened separately and inserted. This is a clockmaker's instrument; you are unlikely to need this tool unless you are making clock parts because it will not accept assemblies of wheel, pinion and arbors.

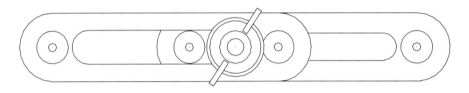

Minimum center distance, about 12mm
Maximum center distance, about 50mm with a pinion dia. of 12mm

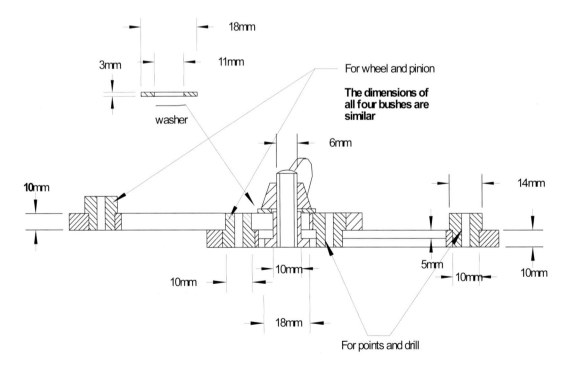

Fig. 11.01   Flat depthing tool.

Fig. 11.01 shows the assembled tool. It consists of two plates that are slotted and bolted together. Each plate has two holes; one of each pair accepts a gear (wheel or pinion) and the other holds a point for marking or alternatively a drill bush for drilling a pivot hole or locating over an existing hole.

MAKING THE FLAT DEPTHING TOOL

Materials needed:
- Brass:
  25mm × 10mm × 500mm (for the slotted plates)
  12mm diameter × 140mm (for bushes and locking nut)

- Steel:
  Six 4M × 6mm countersunk screws
  25mm × 6mm [x] 250mm long BDMS (bright drawn mild steel), or free cutting brass of the same size
  One 6M × 50mm screwed rod, 'allthread'
  One 6mm wingnut (thumb nut) and washer
  Four 4M locking or grub screws

- A selection of drill rod pieces to match the diameters of the arbors that will be set in the tool
- A piece of flat steel to silver solder onto the brass locking nut (3mm [x] 18mm [x] 16m)

Tools needed:
- Twist drills
- 0.5in (12.5mm) end mill
- Hold down' straps. Odd ends of rectangular BDMS drilled to suit the central bolt of the tool post

Stage 1: Preparing the Plates

1. Rough out the plates, cutting to length and marking the slot. It is important that the hole centres should be identical, so the plates should be firmly clamped together before drilling.
    The lower plate has a slot that is rebated so that the tool can be clamped onto clock plates and the pivot holes drilled through it. This rebate will then prevent the square head of the locking screw from turning.

Stage 2: Joining the Plates

2. When both plates have been filed to overall length and width, clamp them together and drill the two end holes (Fig. 11.03). These holes are shown as 10mm in diameter; they must be parallel-sided and dead square to the surface of the plate. Do not go straight in with the full-sized drill; make a smaller hole first and then open up and ream out to the finished size. If you do not have engineering reamers, use a clockmaker's broach and take it through to the parallel section

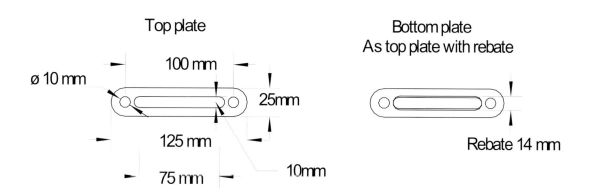

Fig. 11.02   *Rough out the two plates, cutting to length and marking the slot.*

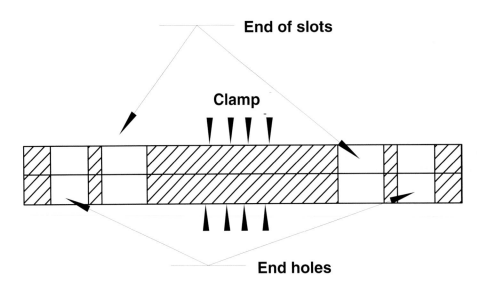

Fig. 11.03   When both plates have been filed to size, clamp them together and drill the two end holes.

at the top. This is a very good reason for making the depthing tool out of brass: steel is not easy to broach.

Two holes to make the ends of the slot will ease the task of milling on the lathe.

The diameter is not itself important so long as it remains truly cylindrical from one plate to the other – the inserts can be machined to suit the holes.

3. Use the holes to bolt the plates together, using bolts with a turned section that close-ly fits the holes (fitted bolts).

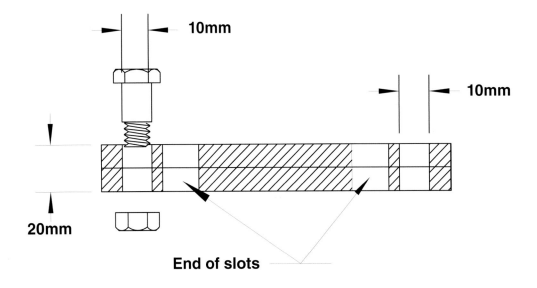

Fig. 11.04   Two holes to make the ends of the slot will make milling easier later.

4. Now file all over, rounding the ends, and creating a solid block of metal whose sides are dead square – test with an engineer's square (Fig. 11.05).

Stage 3: Milling out the Slot and Rebate

5. This block can now be clamped to the cross slide of the lathe for milling out the slot (Fig. 11.06) and the rebate on one side (Fig. 11.07). Note that I have placed two bars of steel immediately on the top edge of the plates; this is so that I can adjust the screws until the pressure is equal on both plates. Testing the tightness of each bar will prove whether one bears down heavier or lighter than the other.

6. The rebate (14mm wide) does not have to be made precisely, but the sides ought to be parallel within a tolerance of about 0.2mm to prevent the square head of the locking screw from twisting when it is at one end or the other. Errors of centring the rebate on the slot are not important either, as the square head can be made to suit.

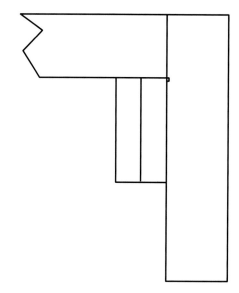

Fig. 11.05   File all over, rounding the ends and creating a solid block of metal whose sides are dead square. Test with an engineer's square.

Packing pieces

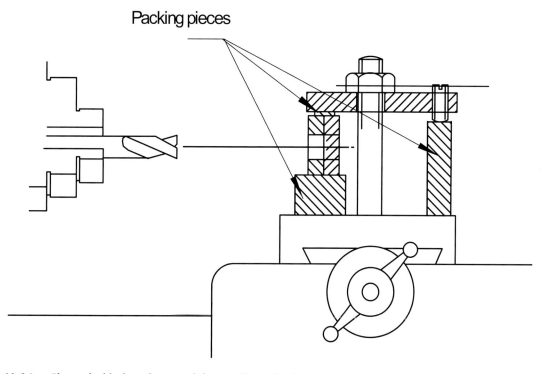

Fig. 11.06   Clamp the block to the cross slide to mill out the slot.

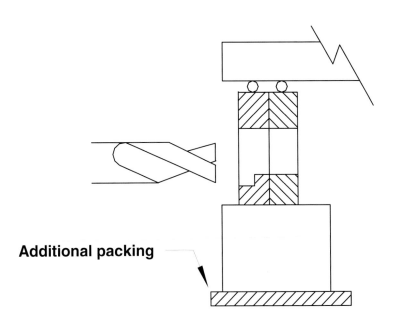

Fig. 11.07   Mill out the rebate on one side.

Additional packing

An end mill with a parallel shank will do this job very well, and can be easily obtained from tool merchants. Use one that has a diameter equal to the width of the slot, and hold it in the three-jaw chuck. Top speed of the chuck when cutting steel is about 450rpm and, for machining quality brass, 1,500rpm. Pack the block so that the centre of the end mill coincides with the centre line of the slot. Make sure that there is no lift in the headstock, otherwise the end mill will climb upwards when winding the cross slide inwards and pull down again on the outward traverse. This will result in a slot that is not smooth-sided.

You must also tighten up the cross slide if the traverse screw has any backlash in it. This is not so important now while you are cutting the through slot because the mill bears on top and bottom of the slot, but when it comes to cutting the rebate with the same size of cutter it is important.

If the backlash is too great and you cannot stop the slide being pulled inwards by the mill, you must cut only during the outward movement. Only move the slide inwards when there is no cut on the tool. You may even find, if the lathe is very sloppy, that you need to stop the rotation of the chuck while traversing back for the next

cut. When you reach the end of the slot, reverse the motion of the traversing handle to remove the backlash before taking more cut.

Use the divisions on the traverse dial to mark the starting and stopping places of the slot. If you have no division dial, mark the apron and the handle boss with chalk.

7. When the slot is right through the two plates, you can cut the rebate, but first you should allow the cutter to dwell at each end of the slot to remove the multiple surfaces produced by the progression of the slot. If you have another end mill that is wide enough to cut the rebate directly, you can go ahead without altering the set-up, but do remember that your marks for each end of the slot will be different with a larger mill.

Stage 4: Finishing the Guides

8. Remove the block from the lathe and clean it up with emery paper to get rid of all fraze from slotting. Do not part the plates yet.
9. The side guides are made out of steel strip. Rough them out and then drill through to produce tapping holes in the top plate (Fig. 11.08).
10. Open up the holes in the guides and countersink them.

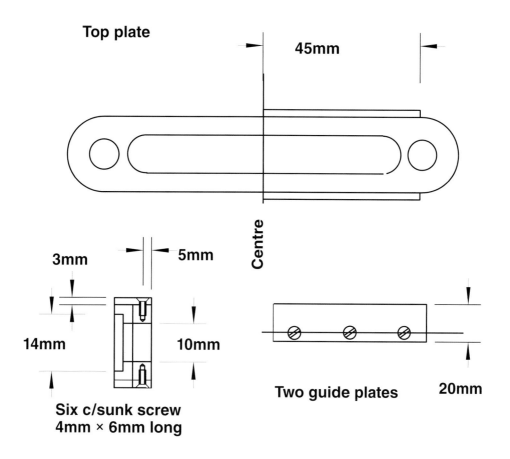

Top plate

45mm

Centre

3mm

5mm

14mm

10mm

20mm

Two guide plates

**Six c/sunk screw
4mm × 6mm long**

Fig. 11.08   Drilling through the side guides to produce tapping holes in the top plate.

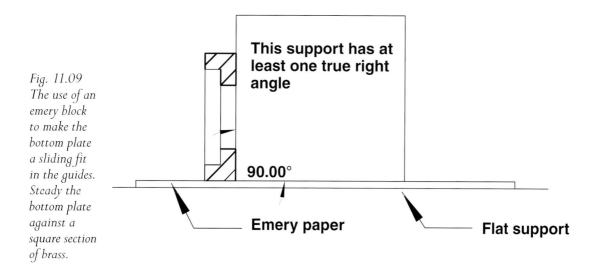

*Fig. 11.09
The use of an
emery block
to make the
bottom plate
a sliding fit
in the guides.
Steady the
bottom plate
against a
square section
of brass.*

**This support has at
least one true right
angle**

90.00°

Emery paper

Flat support

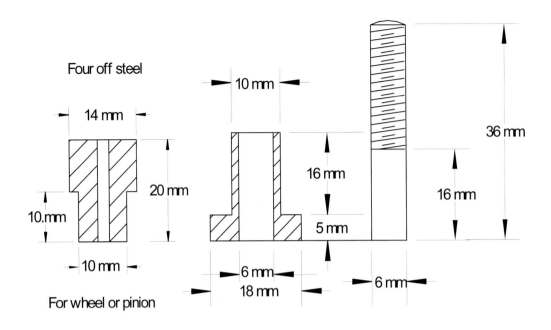

Four off steel

14 mm

20 mm

10.mm

10 mm

For wheel or pinion

10 mm

16 mm

5 mm

6 mm

18 mm

36 mm

16 mm

6 mm

Fig. 11.10   These are the bushes or guides for the markers or drills.

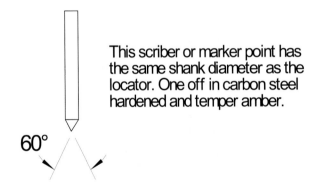

This scriber or marker point has the same shank diameter as the locator. One off in carbon steel hardened and temper amber.

60°

Harden both and temper to amber.

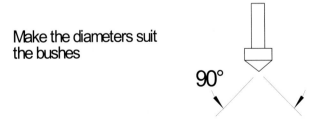

Make the diameters suit the bushes

90°

Locator point, one off in high carbon

steel, hardened and tempered.

Fig. 11.11   The inserts and the points should be hardened and tempered light straw.

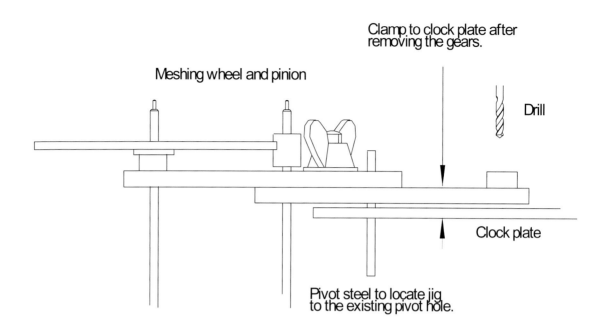

Clamp to clock plate after removing the gears.

Meshing wheel and pinion

Drill

Clock plate

Pivot steel to locate jig to the existing pivot hole.

Fig. 11.12   The tool used to guide a drill while locating on an existing hole.

11. Remove all fraze from their underside and then use steel screws to fasten them to the plate. Now you can undo the fitted bolts that hold the plates together; you should find that the guides clasp the plate so tightly that you need to loosen a few screws to get it out.

12. Steady the bottom plate against a square section of brass (Fig. 11.09) and rub it over a board with fine grades of emery paper on it. Successively emery with 120, 300, 600 and 1,000 grit paper, until when the guides are screwed tight again, the loose plate slides nicely between them without wobbling.

13. The main part of the tool is now complete; turn the inserts for the arbors or points and drill for the locking screw. All the bushes have 'heads' of 14mm diameter to increase the rigidity of the tool. Do not part them off the bar until you have tested them in the slotted plates and seen that they slide smoothly in the slot.

14. I would make the bushes a good press fit, but if you like they can have a flat filed on the 10mm diameter and have the plates tapped for grub screws to engage and lock the bushes. Fig. 11.10 also shows the T nut for locking the two parts of the tool together. The 25mm length of threaded rod can be Loctited into the nut after degreasing thoroughly. Alternatively, the steel plate (15mm [x] 14mm) may be silver soldered onto the rod. Be careful to keep the screwed rod perpendicular to the rectangular steel plate.

15. The inserts and the points should be hardened and tempered light straw (Fig. 11.11).

As you can see, the tool has two pairs of holes with precisely the same centres as each other, and one pair mimics the other as the pieces slide to obtain a proper mesh for two gears. When mesh (or depth) has been achieved the slides are locked together.

Fig. 11.12 shows the tool used to guide a drill while locating on an existing hole, while in Fig. 11.13 the more usual method of striking arcs from two existing holes is employed.

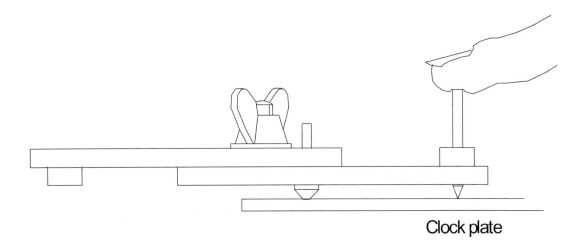

Clock plate

Fig. 11.13   Striking arcs from two existing holes. Remove the wheel and pinion after finding the correct depthing.

Chapter 12

Centre-Marking Tools

A simple centre-marking tool is based on a very old one called a 'priest' or 'preacher' (I do not know why) and, although it again makes little use of a lathe, it is so useful (and I believe is not available commercially) that I have included it. Basically it measures or sets the distance between centres of holes using a vernier calliper. I would expect it to be accurate to about 0.1mm, which is twice the error that can be expected from a vernier regardless of whether it has a dial readout or not.

Fig. 12.01 is a view of the top of the instrument while Fig. 12.02 shows the underside. The distance between points when opened to 60 degrees is 75mm. The points have flats parallel to and aligned with their centre lines; measuring across these flats will produce the distance between the points, which can be used to scribe arcs in the clock plate or to locate in holes or previously centre-punched marks. The accuracy depends on two factors:

- The sensitivity with which the vernier is used
- The rigidity of the tool

Fig. 12.01   Centre-marking tool (view from top).

Fig. 12.02   Centre-marking tool (underside view).

I have shown the distance between the plates that form each arm as 10mm. Support and rigidity would be greater if this distance was 20mm but the tool would be clumsier to use – a classic swings and roundabouts situation.

Fig. 12.03 is a sketch of the instrument. It is made from two pairs of arms cut from 2mm-thick hard brass, pivoted together and carrying marking points at the outer ends. These points have a flat that precisely divides them in two at the top and enables measurements to be made with a vernier calliper.

To use the instrument, the pivot clamp (wing nut) is loosened and the arms moved until the flats on the top of the points are pressed against the inside of the vernier jaws and set to the proper centre distance. The pivot clamp is then tightened and the distance over the points checked again. By laying one point in the centre-punched mark for the first wheel, an arc can now be drawn for the centre arbor pivot, a light tap given on the other point with a small brass hammer where this arc crosses

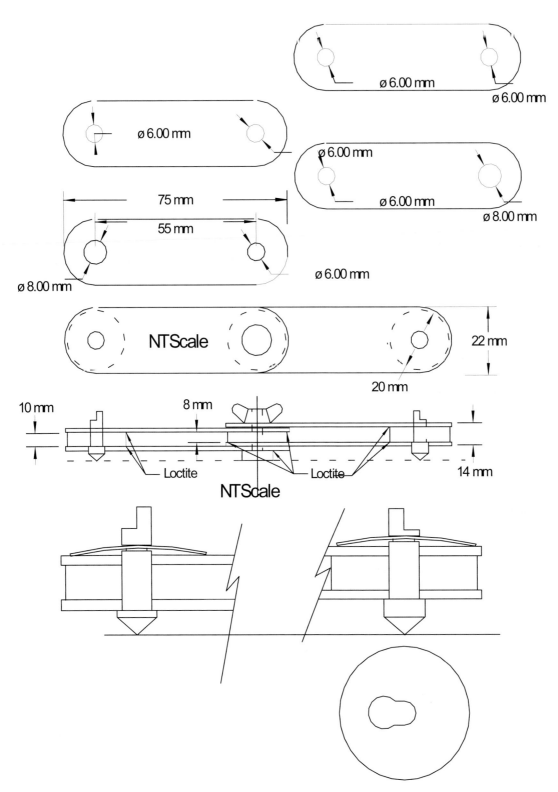

Fig. 12.03   A sketch of the centre-marking tool.

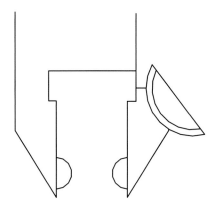

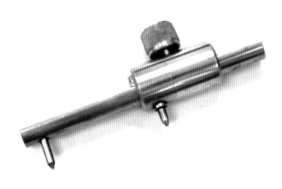

Fig. 12.04   Using a vernier calliper with the centre-marking tool.

Fig. 12.05   A more complicated centre-marking tool with the points carried in the body and a slide.

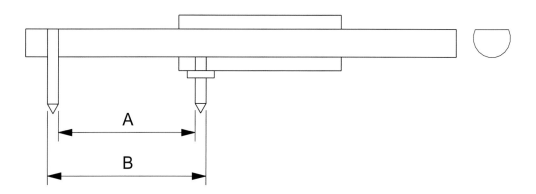

A

B

The distance between centers is found by adding A and B and dividing the sum by 2.

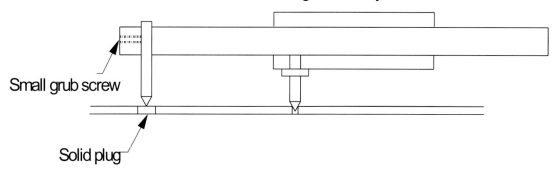

Small grub screw

Solid plug

Fig. 12.06   A sketch of the version with the slide.

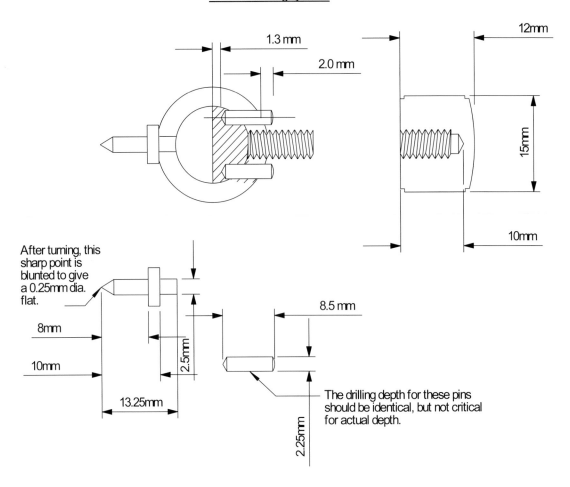

Detail of locking system.

After turning, this sharp point is blunted to give a 0.25mm dia. flat.

The drilling depth for these pins should be identical, but not critical for actual depth.

Fig. 12.07   Diagram showing the method of keeping the slide properly aligned.

the vertical centre line and the position of the centre pivot marked securely. Only one of the points is actually sharp; the one that is placed in a centre-punch mark has a small flat at the apex so that it registers on the conical sides of the centre-punch mark and not the blunt bottom of the mark.

Another instrument is shown in Fig. 12.05. This is more difficult to make but is more accurate because the points are carried in the body and a slide. The first instrument has a tendency for the points to wobble very slightly between the two elements of the arm. Fig.12.06 is a sketch of the version with a slide; and Fig.12.07 shows the method of keeping the slide properly aligned and sliding easily. All but the flat on the sliding bar make use of techniques already detailed, and the flat can be machined by holding on spacers and using a milling cutter in the chuck. Straps will be needed, preferably with a 'V' cut underneath, and the steel bar should be longer than the finished product to allow for the straps. Fig.12.08 suggests a set-up.

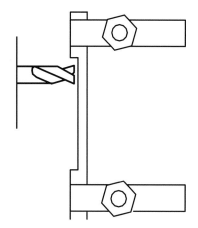

Packing pieces

Adjustment screw

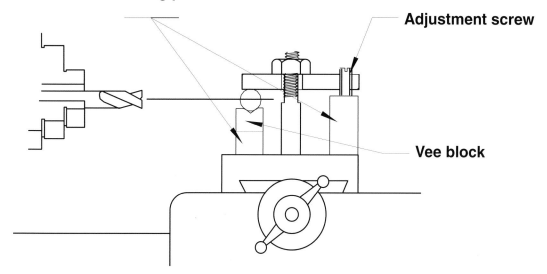

Vee block

Fig. 12.08   A set-up using the centre-marking tool.

Epilogue

There is a great deal more to learn about the use of a lathe but my intention here was to provide a broad range of basic information that would cover a large number of the tasks that a clockmaker, repairer, restorer or model-maker needs.

Only one thing more is needed to make the engineer's or centre lathe a universal machine, and that is a column mounted on the bed, which adds a third dimension to the machining capabilities. Many lathe manufacturers supply a column or a vertical slide. From that point, a machinist can add a rotating head and computer-aided machining (CAM) software and the articles that can be made on it are then only limited by size and the machinist's skill.

Given a lathe, a skilled machinist can even make a larger lathe with three-dimensional

Glossary

Arbor   A spindle that carries a clock wheel or pinion.

Broaching   Increasing the size of a hole with a hand tool. A cutting broach has at least one cutting edge along the length.

Broaching   Polishing broach has no cutting edge, but longitudinal streaks made with emery paper. From smooth to rough.

Burnishing   Producing a mirror-like surface by displacing the metal with a tool.

Carbide   A shortening of tungsten carbide, a hard material formed into cutting tools.

Carborundum   An abrasive material produced by burning sand (silica).

Chatter   Vibration of the cutting tool.

CNC   Computer Numerical Control: controlled by computer.

Emery   A very hard aluminium compound (alumina).

End mill   A cylindrical cutting tool that cuts on the end as well as the circumference.

Form cutting   Using a shaped tool to produce a shaped product such as a clock pillar.

Gap   The space between the main slide (bed) and the headstock.

Going barrel   The container for the spring driving the movement.

Interrupted cut   The intermittent cut that results from having spaces around the periphery of the work.

Lead screw   A screwed rod that moves the saddle of a lathe along the bed.

Mandrel   A spindle or shaft; the main spindle of a lathe.

Module   In gear formulae this is the ratio of number of teeth to pitch diameter.

Movement   Name for the 'works'

Pillar   A supporting post for the clock movement plates, may be round or rectangular (Birdcage movement).

Pinion   A gear with fewer than about sixteen teeth (not a hard-and-fast definition).

Pitch circle   The circle that contact is made between two gears or a gear and pinion.

Sinter   Producing a solid form by heating and pressing a powder.

Slot drill   A cylindrical tool that cuts on the circumference only.

Spring barrel   Alternative name for going barrel.

Swaging   Riveting particularly using a lathe.

Wheel   A gear with more than about sixteen teeth (not a hard-and-fast definition).

Index